Smithsonian Institution

GUIDE TO THE *wildlife* OF
SOUTHWEST CHINA

William J. McShea // Sheng Li // Xiaoli Shen // Fang Wang // Dajun Wang

A Smithsonian Contribution to Knowledge

Smithsonian Institution
Scholarly Press

WASHINGTON, D.C.
2018

Published by
SMITHSONIAN INSTITUTION SCHOLARLY PRESS
P.O. Box 37012, MRC 957
Washington, D.C. 20013-7012
https://scholarlypress.si.edu

Research for this book was funded in part by the Hong Kong Ocean Park
Conservation Foundation.

Cover images: golden pheasant (top left), photo by Li Sheng; golden snub-nosed
monkey (top right), photo by Wang Fang; leopard (bottom left), photo by Li Sheng;
giant panda (bottom right), photo by PKU/eMammal.

Library of Congress Cataloging-in-Publication Data:
Names: McShea, William J., author. | Smithsonian Institution Scholarly Press,
    publisher. | Smithsonian Institution.
Title: Guide to the wildlife of southwest China / William J. McShea [and four
    others].
Other titles: Smithsonian contribution to knowledge.
Description: Washington, D.C. : Smithsonian Institution Scholarly Press, 2018.
    | Series: A Smithsonian contribution to knowledge | Copyright 2018 by
    Smithsonian Institution. | Includes bibliographical references and index.
Identifiers: LCCN 2017043144| ISBN 9781944466138 (paperback) | ISBN
    9781944466145 (ebk.)
Subjects: LCSH: Mammals--China, Southwest--Identification. | Birds--China,
    Southwest--Identification.
Classification: LCC QL729.C5 M37 2018 | DDC 599.0951/3--dc23 | SUDOC SI
    1.60:W 64
LC record available at https://lccn.loc.gov/2017043144

ISBN-13: 978-1-944466-13-8 (print)
ISBN-13: 978-1-944466-14-5 (ebook)

Printed in the United States of America

♾ The paper used in this publication meets the minimum requirements of
the American National Standard for Permanence of Paper for Printed Library
Materials Z39.48–1992.

CONTENTS

MAMMALS OF SOUTHWEST CHINA

Primates

Canids

Bears

Procyonid

Badgers, Martens, Otters, and Weasels

BIRDS OF SOUTHWEST CHINA

COMMON DOMESTIC ANIMALS AND OTHER WILDLIFE

INTRODUCTION

The purpose of this field guide is to inform reserve staff, students of natural history, and the casual tourist of the wildlife present within the reserves of southwest China. This English edition will, ideally, be followed by a Chinese language edition to inform as diverse an audience as possible. The original reserve staff employed in many reserves were local villagers, and they had a broad knowledge of the local wildlife and their habits. As the Chinese reserve system matures, more university students from urban areas are being assigned to work in reserves, and more staff are being transferred between localities. With the retirement, relocation, and promotion of the original field staff, a primary knowledge base of wildlife sign identification is being lost. One goal of this guide is to help capture as much of this knowledge as possible before it is lost. In some cases, descriptions of natural history rely on local knowledge or published references for some of the species.

With the exception of diurnal mammals on the open plateau and the largest animals within the forest, many species are not readily observable. Camera-trap surveys prove helpful in this regard, but some animals' secretive habits allow even the most dedicated naturalist only rare glimpses. What is more obvious are the signs left behind by these animals—if you know what to look for. Subtle differences in feces, tracks, and feeding sites reveal the diversity within the forests of southwest China. Until these animals become more habituated to human presence, tracks and other signs are the best way to appreciate and catalog Chinese wildlife.

Photographs

The authors are ecologists with a long history of conducting field surveys in the region, recently through the use of wildlife cameras (also known as trip cameras, camera traps, or trail cameras). These camera units are commercially available in most developed and developing countries, including China. The quality of the products may vary, but the basic unit is a heat-motion sensor connected to an autofocus camera. To allow nighttime photographs, a light sensor activates an infrared flash when needed. Programming allows one to set the parameters for number of photographs per sensor detection, active time of day, and sensor sensitivity. The unit is encased in waterproof housing and can easily be attached to a tree or post. The result is a unit that can detect all large-bodied (>500 g) terrestrial species moving within a 1- to 30-m window in front of the unit (Figure 1). However, the cameras are not perfect; we have very few photographs of arboreal mammals, which spend the majority of their time well above the range of the camera sensor. Other limitations include the following: (1) The sensor units are not as sensitive in tropical climes as they are in temperate (cooler) ones. (2) The cameras and sensors are not as durable or sensitive in rainy locales as they are at drier sites. (3) The sensor range does not always match the range of the infrared flash, resulting in "false" triggers during the night. (4) Finally, the distance the sensor monitors depends on the density of vegetation,

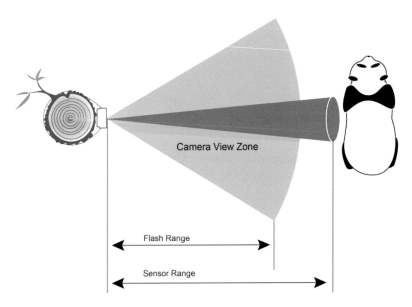

Figure 1. Typical operation of a wildlife camera used to detect many mammals in this guide. A motion-heat sensor is attached to a camera unit that can autofocus on the animal and utilize an infrared flash if needed. Illustration by Yu Wei.

the ambient temperature, and the size of the animal. Despite all these caveats, game cameras have revolutionized the survey of mammals and large birds in natural areas worldwide.

Our original intent for this guide was to include trail camera photographs of each wildlife species, obtained during surveys. In reality, some of the arboreal species have not been captured with trail cameras, and some of the more tropical habitats have not been completely sampled. In those cases, it was necessary to obtain photos from naturalists or, occasionally, zoos. The best photographs available for each species were selected, with the intent of displaying the unique identifying marks of each species. For some species, nocturnal or secretive habits preclude a clear photograph showing critical identifying features. In these cases, the chosen camera-trap photographs still show these animals moving through their natural habitats. The photographs have limitations, but most of the species will rarely be encountered in the wild, and their images provide a sense of the essence of these animals.

Both professionals and casual hikers are likely to encounter feces, tracks, and other signs of these animals. Images from natural settings containing common items are used as a reference of scale. When tracks or other signs are not readily observed, as with many arboreal mammals, species have been excluded from the guide or photos of captive animals and their signs have been used. When tracks and other signs of closely related species are difficult to distinguish, the species are grouped, and a single example for the group is provided.

Geographic Range and Physiography of the Region

Many significant decisions were made regarding the spatial extent of this field guide. The original intent was to create a guide to wildlife within the reserves for giant pandas (*Ailuropoda melanoleuca*). However, one conservation value of this guide is to inform reserve staff, and we find giant panda reserve staff are the best trained within China. An urgent need exists to provide training and knowledge to staff outside of giant panda reserves, where funding levels are generally lower and fewer opportunities for training are available. With the optimistic attitude that this guide will find its way into the hands of reserve staff outside of the giant panda reserves, this guide covers all of Sichuan, Yunnan, and Guizhou Provinces; Chongqing Municipality (hereafter referred to as Chongqing City) and Guangxi AR

Figure 2. The extent of China covered by this field guide.

Figure 3. (*opposite*, top) National and Provincial Nature Reserves present within the region as of January 2015.

Figure 4. (*opposite*, bottom) Elevation range found within the guide's extent with the six major rivers that flow through the region.

(Autonomous Region); the eastern regions of Qinghai Province and Tibet AR; and the southern portions of Gansu and Shaanxi Provinces (Figure 2).

The focus of protected reserves in the region has been on China Protected List I species—the highest level of concern—including giant pandas, Przewalski's gazelles (*Procapra przewalskii*), takin (*Budorcas taxicolor*), and golden snub-nosed monkeys (*Rhinopithecus roxellana*; Figure 3). Large expanses of the plateau (both Qinghai and Tibet) have protected status, as does western Sichuan Province. Southern Gansu and Shaanxi Provinces also contain significant giant panda reserves. Reserves in the remaining provinces are smaller and centered on either pheasant or langur species. The exception is Yunnan Province, with golden monkey and additional reserves along the border with Southeast Asian countries for remnant populations of elephant (*Elephas maximus*), gibbons, and langurs.

The extent of this region is approximately 2,400,000 km² and ranges from sea level in the southeast to 7,590 m elevation in the west (Figure 4).

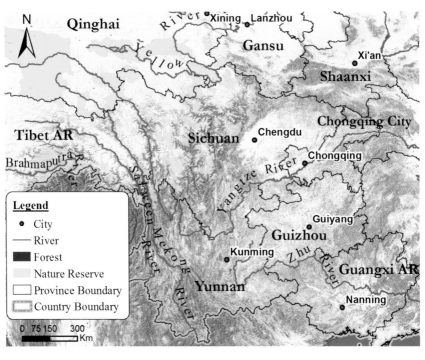

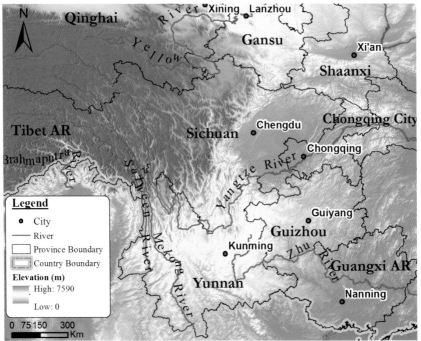

It includes a major portion of six major river drainages: Yellow, Yangtze, Mekong, Salween, Brahmaputra, and Zhu (Pearl) Rivers. All but the Zhu River flow from the Tibetan Plateau into the lowlands of the east and south.

The plateau (Tibet AR and Qinghai Province) is characterized by rolling grasslands with mountains and outcroppings and coniferous trees distributed along river valleys. The transition area from the plateau to the lowlands (Sichuan and Gansu Provinces) is characterized by steep slopes

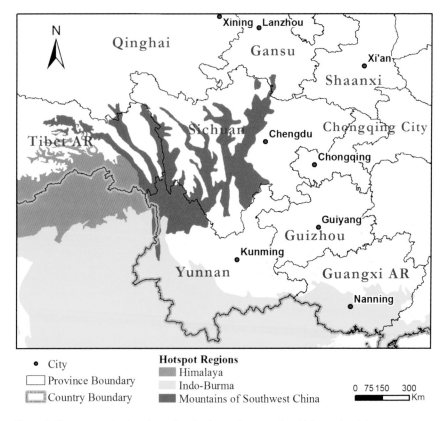

Figure 5. Three recognized regions of endemic species, as well as high species richness and biodiversity, are found within the area covered in this guide. Southwest China has the convergence of the Indo-Burma and Himalaya regions, as well as the complex topography of the mountains of southwest China. The map is derived from the biodiversity hot spot data set maintained by the Critical Ecosystem Partnership Fund (http://www.cepf.net/resources/hotspots/Pages/default.aspx).

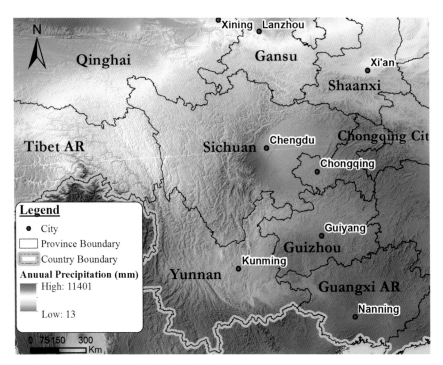

Figure 6. Annual rainfall (in mm) for the region, obtained from the WorldClim data set (http://www.worldclim.org/).

with narrow valleys and ridges. Along the southern and southeastern edges of the plateau (Yunnan Province) the terrain is more variable, especially along the border with Laos, Vietnam, and Myanmar. Outside of these regions, agriculture and human developments begin to dominate the landscape, with lower mountain ranges (Shaanxi Province and Chongqing City) or limestone outcroppings (Guangxi AR and Guizhou Province) containing most of the remaining forest. Increasingly, accessible forest around human developments is being converted to planted forests, except where nature reserves have been established. Most of the unique biodiversity of the region is located in the complex topography of western Sichuan Province and eastern Tibet AR, as well as the border with Southeast Asia (Figure 5).

Annual rainfall varies significantly across the region, with the plateau region being essentially a high-elevation desert (13 cm annually) and the southern extent being a tropical rainforest (>800 cm annually; Figure 6). In general, the annual rainfall increases from the northwest to southeast within the region.

Figure 7. The density of humans (mean number per square kilometer) across the region. The plateau region has very low densities, and the lowlands of the Sichuan basin are heavily populated. Data source: Gridded Population of the World, Version 3, downloaded from NASA's Socioeconomic Data and Applications Center, http://dx.doi.org/10.7927/H4XK8CG2.

As critical to the region's wildlife as the topography and rainfall is the human population density. Humans are not evenly distributed (Figure 7). A general trend is increased human density from west to east, but the highest population densities occur throughout the lower elevations of the Sichuan basin in the eastern portions of Sichuan Province. Within the range of this guide, all of the plateau (Tibet AR, Qinhai Province, and western Sichuan, Gansu, and Yunnan Provinces) has very low human densities. In general, the provincial capital serves as a focal area of human development within each province (Figure 7).

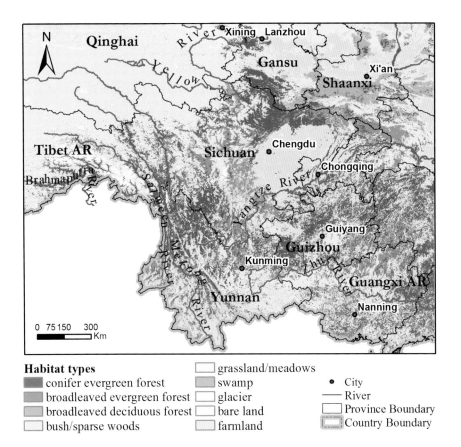

Habitat types

- conifer evergreen forest
- broadleaved evergreen forest
- broadleaved deciduous forest
- bush/sparse woods
- grassland/meadows
- swamp
- glacier
- bare land
- farmland
- • City
- —— River
- Province Boundary
- Country Boundary

Figure 8. Habitat types found within the extent of the guide as derived from Global Land Cover 2000 by the Joint Research Centre of the European Commission.

It is not possible to provide an in-depth explanation of the habitat types found within the range of this guide (Figure 8). Three major habitat types are relevant to wildlife: agricultural landscapes (including planted forests), high-elevation grasslands, and mid- and low-elevation forests.

Agricultural Lands

For agricultural landscapes, two general types are applicable to wildlife: crop/forest mosaics in the mountains and expansive croplands (often rice cultivation) in the lowlands. Among forest types there is a third human-modified habitat with planted forests, either coniferous or deciduous.

A mosaic of agriculture and forest. Pingwu County, Sichuan Province

Rice paddies. Medog County, Tibet AR

A monoculture of young planted fir (Abies) trees. Zhaotong City, Yunnan Province

Rubber plantation. Jinghong City, Yunnan Province

Grasslands

The plateau in the western regions of this guide has broad valleys with grassland or a mixture of grass and shrubs in sheltered areas. These broad valleys progress up to rocky crags without vegetation. In the eastern and southeastern parts of the plateau the valleys become steeper, and the lower elevations are populated with conifer forests. In northern areas of the plateau the grasslands can be xeric and considered semidesert. Alpine grasslands are found along the mountains that rim the edge of the plateau. These grasslands are bordered by forests and can be isolated on mountain tops.

High elevation grassland. Sanjiangyuan NR, Qinghai Province

Alpine valley with schree. Bomi County, Tibet AR

High elevation grass/shrubland valley. Sichuan Province

Semidesert grassland common on northern plateau area. Qinghaihu NR, Qinghai Province

Natural Forests

Natural forests include coniferous or evergreen forest in colder climes and mixed deciduous and coniferous in temperate regions and evergreen in semitropical locations. These are gross generalizations as the extreme topography results in rapid changes over short distances. The highest-elevation forests are found in sheltered valleys or southeastern slopes within the plateau and are either coniferous or alpine oak.

Alpine meadows embedded within forested mountains. Batang County, Sichuan Province

Coniferous forest with evergreen understory. Wanglang NR, Sichuan Province

Mixed deciduous and evergreen forest usually at sites with abundant rainfall or fog. Medog County, Tibet AR

Mixed deciduous/coniferous forest with a relatively high diversity of tree species, usually abundant water in narrow streams.

Alpine oak (Quercus) forest at high elevation. Gexigou NR, Sichuan Province

Coniferous forest. Wanglang NR, Sichuan Province

Female wild boar with offspring in young forest opening with bamboo understory. Changqing NR, Shaanxi Province

Tropical forest. Medog County, Tibet AR

Species Selection

For areas along the southern edge of China, past reports contain several tropical species, such as tigers (*Panthera tigris*) and northern white-cheeked gibbons (*Nomascus leucogenys*), that have not been detected in recent years. For such species, their descriptions are included because either good habitat still exists and may be recolonized, or remnant populations may persist undetected. Most species under 500 g are excluded because of their low number of sightings, lack of unique sign under most circumstances, or rarity. Although the black giant squirrel (*Ratufa bicolor*) is included because of its obvious diurnal habits, individual species in the diverse group of other squirrels have been excluded. Strong arguments can be made for other groups, and maybe they will be included in a future edition of this guide. New species and new records are being detected; these will be included in future editions.

Several pheasant and grouse species appear because they are among the more visible species in many reserves, and southwest China is an incredibly diverse region for these birds. The final section, "Common Domestic Animals and Other Wildlife," lists animals that leave obvious, unique signs when present, that are commonly present in reserves, and whose signs can be confused with that of related wildlife.

Using the Guide

The species descriptions are grouped according to their phylogeny, and within each group, the species are arranged alphabetically according to common name. Each species description includes the common name, scientific name (follows nomenclature of IUCN Red List), pinyin name (pronunciation of Chinese characters), and, when applicable, alternative common names.

Information on threatened status and body measurements accompany each species. Conservation indicators are as follows:

IUCN (International Union for Conservation of Nature): risk of extinction from the IUCN Red List; categories are CR (Critically Endangered), EN (Endangered), VU (Vulnerable), NT (Near Threatened), LC (Least Concern), DD (Data Deficient), and NE (Not Evaluated).

CITES (Convention on International Trade in Endangered Species of Wild Fauna and Flora) Checklist: indicates species listing in Appendix I, II, or III or NL (Not Listed). Appendix I is the highest level of concern.

CPS (China Protected Species): class I, II, or NL

CRL (China Species Red List): same as IUCN Red List

The data are displayed the following way:

IUCN: CR, EN, VU, NT, LC, DD, NE

CITES: I, II, III, NL

CPS: I, II, NL

CRL: CR, EN, VU, NT, LC, DD, NE

The conservation status of each species is based on the IUCN Red List (http://www.iucnredlist.org) and Chinese law as of September 2017. A species' status may change with time; thus, it is recommended that readers check the IUCN Red List for current levels.

The body measurements include body length and mass measures. All measures are for adult individuals.

BL (body length): for mammals, from the tip of the nose to the boney tip of the tail; for birds, from the tip of the beak to the end of the tail feather.

TL (tail length): from rump to tip.

WT (weight): when the sexes are markedly different in size, both weights are listed.

The data are displayed the following way:

BL: centimeters (cm) to meters (m)

TL: centimeters (cm)

WT: grams (g) to kilograms (kg)

Species range maps were generated by the IUCN and show the distribution of each mammal across the region. Each map covers the same area and includes the boundaries and the capital of each province for reference. Controversy exists over the boundary between China and India; the English version of this guide recognizes the international convention. A future Chinese version will use the Chinese government boundary map. Where local knowledge differs from the IUCN range maps, alterations have been made and explained. Most of our images are from within the National and Provincial Nature Reserves; when indicating an image location we have identified the specific reserve and abbreviated "NR".

Most species descriptions follow published accounts. Some terms used to describe an animal or track might be unfamiliar to the reader, and drawings are provided with explanations. Most descriptive terms are self-explanatory, but the illustrations below indicate where features are located on animals.

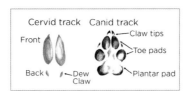

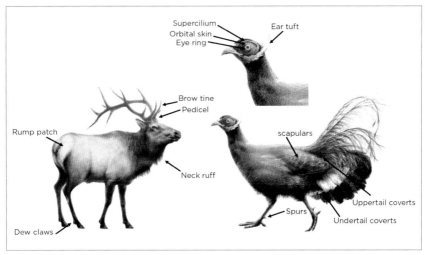

Each distribution map indicates the range of the species in green, but within the range each species has specific habitat associations that limit their distribution. Maps used in this volume were derived from distribution layers obtained from the IUCN Red List of Threatened Species as of summer 2017. Some distributions have been modified based on the authors' local knowledge. A section on ecology is included for each species to help clarify observations of behavior or group size. This information was drawn from the authors' personal experiences, interviews with reserve staff, and published texts (see the bibliography). Each of these references was used repeatedly to develop this field guide, especially Smith and Xie (2008) for the mammals and MacKinnon and Phillipps (2000) and Zheng (2015) for the pheasants. We humbly acknowledge the use of their knowledge in the descriptions for some of the less familiar species.

Bibliography

Bang, P., P. Dahlstrøm, and M. Walters. *Animal Tracks and Signs.* New York: Oxford University Press, 2001.

Castello J. R. *Bovids of the World.* Princeton, NJ: Princeton University Press, 2016.

Elbroch, M. *Mammal Tracks and Sign: A Guide to North American Species.* Mechanicsburg, PA: Stackpole Books, 2003.

Fan, P. F., K. He, X. Chen, A. Ortiz, B. Zhang, C. Zhao, Y. Q. Li, H. B. Zhang, C. Kimock, W. Z. Wang, C. Groves, S. T. Turvey, C. Roos, K. M. Helgen, and X. L. Jiang. Description of a new species of *Hoolock* gibbon (Primates: Hylobatidae) based on integrative taxonomy. *American Journal of Primatology* 79 (2017):e22631.

Feng, Z. J., G. Q. Cai, and C. L. Zheng. *The Mammals of Xizang.* Beijing: Science Press, 1986.

Francis, C. M. *A Guide to the Mammals of Southeast Asia.* Princeton, NJ: Princeton University Press, 2008.

Geissmann, T., N. Lwin, S. S. Aung, T. N. Aung, Z. M. Aung, T. H. Hla, M. Grindley, and F. Momberg. A New Species of Snub-Nosed Monkey, Genus *Rhinopithecus* Milne-Edwards, 1872 (Primates, Colobinae), from Northern Kachin State, Northeastern Myanmar. *American Journal of Primatology* 73 (2011):96–107.

Grimmett, R., C. Inskipp, and T. Inskipp. *Pocket Guide to the Birds of the Indian Subcontinent.* London: Christopher Helm, 2002.

Groves, C., and P. Grubb. *Ungulate Taxonomy.* Baltimore: Johns Hopkins University Press, 2011.

Hunter, L. *Carnivores of the World.* Princeton, NJ: Princeton University Press, 2011.

IUCN. IUCN Red List of Threatened Species. http://www.iucnredlist.org/ (accessed 2 October, 2017).

IUCN Cat Specialist Group. *Cats in China.* Beijing: Forestry Publishing House, 2013.

Jiang, Z. G., J. Jianping, W. Yuezhao, E. Zhang, and Z. Yanyun. Red List of China's Vertebrates. *Biodiversity Science* 24, no. 5 (2016):500–551.

Jiang, Z. G., M. Yong, W. Yi, W. Yingxiang, Z. Kaiya, L. Shaoying, and F. Zuojian, eds. *China's Mammal Diversity and Geographic Distribution.* Beijing: Science Press, 2015.

Jiang, Z. G., S. Y. Liu, Y. Wu, X. L. Jiang, and K. Y. Zhou. China's Mammal Diversity (2nd edition). *Biodiversity Science* 25, no. 8 (2017):886–895.

Li, C., C. Zhao, and P. F. Fan. White Cheeked Macaque (*Macaca leucogenys*): A New Macaque Species from Modog, Southeastern Tibet. *American Journal of Primatology* 77 (2015):753–766.

Ma, S. L., X. F. Ma, and W. Y. Shi. *A Guide to Mammal Tracking in China.* Beijing: China Forestry Publishing House, 2001.

MacKinnon, J., and K. Phillipps. *A Field Guide to the Birds of China.* With F. Q. He. Oxford: Oxford University Press, 2000.

Madge, S., and P. McGowan. *Pheasants, Partridges and Grouse: A Guide to the Pheasants, Partridges, Quails, Grouse, Guineafowl, Buttonquails and Sandgrouse of the World.* London: Christopher Helm, 2010.

Petter, J.-J., and F. Desbordes. *Primates of the World: An Illustrated Guide*. Princeton, NJ: Princeton University Press, 2010.

Rezendes, P. *Tracking and the Art of Seeing: How to Read Animal Tracks and Sign.* 2nd ed. New York: HarperCollins, 1999.

Robson, C. *New Holland Field Guide to the Birds of South-East Asia.* London: New Holland, 2005.

Sheng, H. L., and Z. X. Liu. *The Musk Deer in China*. Shanghai: Shanghai Scientific and Technical Publishers, 2007.

Smith, A. T., and Y. Xie, eds. *A Guide to the Mammals of China*. Princeton, NJ: Princeton University Press, 2008.

Tang, C. Z., ed. *Birds of the Hengduan Mountains Region*. Beijing: Science Press, 1996.

Wang, S., and Y. Xie, eds. *China Species Red List.* Vol. 2, *Vertebrates,* part 2. Beijing: Higher Education Press, 2009.

Zheng, G. M., ed. *Pheasants in China*. Beijing: Higher Education Press, 2015.

Zheng, Z. X. *Fauna Sinica: Aves*. Vol. 4, *Galliformes.* Beijing: Science Press, 1978.

ACKNOWLEDGMENTS

The authors enjoyed creating this text, as we learned much from each other, from the reserve staff we interviewed, and from the texts we consulted. We hope readers have the same experience as they review each species and marvel at the diversity of Chinese wildlife. We thank all the photographers who provided images of animals and sign. We could not have completed this text without your cooperation. We specifically thank Imaging Biodiversity Expedition (IBE), Kadoorie Farm & Botanic Garden–Kadoorie Conservation China (KFBG-KCC), Chinese Felid Conservation Alliance (CFCA), Sichuan Forestry Department, China Bird Tour, and Nature Image China for image contributions and information. We thank Jenny Santiestevan and Olivia Cosby for keeping us, as well as all the images and maps, organized. Your careful eyes for detail were much needed. We thank our institutions, Peking University and Smithsonian Institution, for supporting this effort. Funding from the Smithsonian's National Zoo and Ocean Park Conservation Foundation Hong Kong made the manuscript possible. We thank Pan Wenshi and George Schaller for inspiring several of us to see the natural world. Finally, we thank all the nature reserve staff for patrolling, walking, climbing, fording, protecting, and watching over the animals of southwest China.

MAMMALS OF SOUTHWEST CHINA

PRIMATES // LING ZHANG LEI

Gibbons // Chang Bi Yuan

General Description

Gibbons include up to 19 living species in 4 genera within the family Hylobatidae. The species are found throughout Southeast Asia and parts of Bangladesh, northern India, southern China, and Indonesia. Gibbons differ from great apes (chimpanzees, bonobos, gorillas, orangutans, and humans) in being smaller and exhibiting little sexual dimorphism except for coat color, and they do not construct nests. However, like all apes, gibbons are large primates lacking tails. Gibbons have long hands and feet, with a deep cleft between the first and second digits of their hands and a flexible wrist that allows biaxial movement. Their fur is usually black, gray, or brownish, often with white markings on the hands, feet, and face.

Habitat and Ecology

Gibbons are exclusively confined to forests and depend on a contiguous canopy to move between trees. They are found in broadleaf, moist deciduous forest; mixed evergreen-deciduous forest with tall deciduous trees and evergreen understory; and subtropical broadleaf forests. In China, they occur across a wide range of elevations from sea level to 2,700 m, including limestone outcroppings, as long as there is forest cover.

Scat

Gibbon feces are loose and smooth. The dark brown-greenish pellets (5 × 1 cm) are clustered but often form separate fecal pellets after falling from the trees.

Tracks

Gibbons live primarily in trees but use a bipedal walk on their hind legs when traversing the ground; when bipedal, they land on their toes before the heel touches the ground, leaving a distinctive five-toe mark. Their tracks (15 × 4 cm) show four toes close together, with a separated thumb pointing at a right angle. Gibbons have larger tracks and a longer thumb print than those of sympatric langurs and leaf monkeys.

Other Sign
Each morning upon awakening, a gibbon troop loudly announces its presence using a territorial hooting call. The call might be responded to by nearby family groups. This noisy display lasts approximately 30–90 minutes.

Black-Crested Gibbon
Nomascus concolor　Hei Guan Chang Bi Yuan

Adult male

IUCN: CR
CITES: I
CPS: I
CRL: CR

BL: 43–54 cm
TL: not applicable
WT: 7–10 kg

Distribution
The species occurs discontinuously in southwest China, northwest Lao People's Democratic Republic, and northern Vietnam. Three subspecies are recognized in Yunnan Province: two subspecies in southwest Yunnan Province, *Nomascus concolor concolor* and *Nomascus concolor furvogaster*, and *Nomascus concolor jingdongensis*, which occurs in west central Yunnan in a small region around

Adult female

Wuliang Mountain, between the Mekong and Chuanhe Rivers. An isolated population occurs in southern Guangxi AR along the China-Vietnam border, which is now considered a separate species *N. nasutus* (Cao-vit gibbon or eastern black-crested gibbon).

Appearance

The genders differ in coat color; the male is almost completely black, with occasional white or buff cheeks, whereas the female is a golden or buff color with variable black patches, including a black streak on the head.

Eastern Hoolock Gibbon
Hoolock leuconedys // Bai Mei Chang Bi Yuan

Adult male

IUCN: VU
CITES: I
CPS: I
CRL: CR

BL: 60–90 cm
TL: not applicable
WT: 6–8.5 kg

Distribution

This species is found in southern China (western Yunnan) and northeast Myanmar (east of the Chindwin River). In China, there are an estimated 50–300 individuals found in Yunnan Province as far east as the Salween River. The species lives in tropical monsoon forests up to 2,700 m in elevation. Recent literature recognizes the populations in western Yunnan Province as a distinct species, the Gaoligong hoolock gibbon *H. tianxing*.

Appearance

Hoolock gibbons are the second largest gibbons in the family Hylobatidae. The males and females are about the same size, but they differ considerably in coloration: males are black colored with a white brow, whereas females have a gray-brown fur, which is darker at the chest and neck. Females have white rings around their eyes and mouth.

Adult female (right) and subadult (left)

Habitat

Scat

Scat

Feeding sign

Northern White-Cheeked Gibbon

Nomascus leucogenys // Bai Jia Chang Bi Yuan

Adult male

IUCN: CR
CITES: I
CPS: I
CRL: CR

BL: 65–90 cm
TL: not applicable
WT: 7–9 kg

Distribution

The northern white-cheeked gibbon may still occur in Xishuangbanna in southernmost Yunnan Province, just across the border from Laos. However, it was last reported in the wild in 2008.

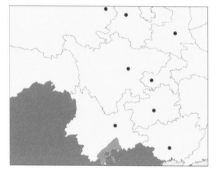

Appearance

The northern white-cheeked gibbon is a large-bodied gibbon. It is sexually dimorphic, with males and females having different coloration and males also being slightly larger. Males have black hair over their entire bodies, except for distinct white patches on their cheeks, as well as a prominent tuft of hair on the crown of the head and a gular sac. Females are reddish tan in color, lack a cranial tuft, and have a crest of black or dark brown fur running from the crown to the nape of the neck.

Langurs and Leaf Monkeys
Ye Hou

General Description
Langurs and leaf monkeys are common names for Asian monkeys belonging to the family Cercopithecidae. Most species are restricted to South and Southeast Asia and are distinguished by their vocalizations and coat color. They are gregarious, diurnal, and mostly arboreal monkeys with long tails and slender bodies. The limbs, hands, and feet are also long and slender.

Habitat and Ecology
Leaf monkeys occur in several habitat types, including primary and secondary forest, agricultural and urban lands, and even mangrove swamps. Regardless of their habitat, they feed mainly on leaves but are also known to consume fruits, grass, and flowers.

Scat
The brownish scats (4 × 2 cm) are conical shaped but may vary with diet. Usually, the scat is composed of leaf segments but may contain seeds and other vegetation.

Tracks
Langurs and leaf monkeys are both semiarboreal and move on all four limbs while on the ground. They leave long and narrow footprints (15–20 cm in length, 3–6 cm in width).

Other Sign
These species live in groups, and both male and female monkeys use vocalizations for alarm, predator detection, group cohesion, and defense. For some species (e.g., white-headed langur), sleeping sites are easy to distinguish because of the brownish urine stains below their limestone roosts (see photos for white-headed langur).

François' Langur

Trachypithecus francoisi // Hei Ye Hou

Adult female and two young

IUCN: EN
CITES: II
CPS: I
CRL: EN

BL: 48–64 cm
TL: 70–90 cm
WT: 5.5–7.5 kg

Distribution

This species occurs in southern China and Vietnam. In China, the isolated populations are distributed in semi-tropical monsoon and moist tropical and subtropical rainforests in limestone areas in Chongqing City, Guizhou Province, and Guangxi AR. The species occurs at elevations from 400 to 1,500 m.

Appearance

François' langur is a medium-sized primate with black silky hair. It has very distinct white sideburns that grow down from its ears to the corners of its cheeks. Hairs on the head form a crest. This species shows sexual dimorphism in its size. Males have a larger body size, with longer tails, than do females. The coat of young langurs is bright orange.

Nepal Gray Langur

Semnopithecus schistaceus // Chang Wei Ye Hou

Adult

IUCN: LC	BL: 62–79 cm
CITES: I	TL: 69–103 cm
CPS: I	WT: 8–24 kg
CRL: CR	

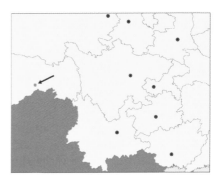

Distribution

The Nepal gray langur is found in the Tibetan AR in Bo Qu, Ji Long Zang Bu, and the Chumbi valleys. This species inhabits subtropical to temperate broadleaf forest, pine forest, montane forest, riverine forest, and rocky outcrops. It can be found up to 3,500 m eleva-tion. The range shown is increased from that indicated by IUCN on the basis of our knowledge of sightings in Mêdog County, Tibet AR.

Appearance

The Nepal gray langur is a gray-colored leaf monkey; the fluffed fur around its head is remarkably white, with black-colored face and ears. There are significant gender differences in size, with the adult male always larger than the female.

Shortridge's Langur

Trachypithecus shortridgei // Dai Mao Ye Hou

Adult females

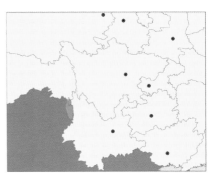

IUCN: EN
CITES: I
CPS: I
CRL: CR

BL: 109–160 cm
TL: 70–120 cm
WT: 9–14 kg

Distribution

This species occurs in evergreen and semievergreen forests in Dulongjiang Valley, Gongshan District, in northwest Yunnan Province and is distributed from 200 to 2,500 m elevation.

Appearance

Shortridge's langur is a large primate with gray pelage. The pelage color is silver gray on the torso and inside of each limb and dark gray on the outside of each limb and on the tail. The species' face has a distinctive fringe of upright hairs with black facial skin and orange eyes.

White-Headed Langur

Trachypithecus poliocephalus // Bai Tou Ye Hou

Two females and a newborn

IUCN: CR	BL: 52–71 cm
CITES: II	TL: 70–90 cm
CPS: I	WT: 6–9.5 kg
CRL: CR	

Distribution

The white-headed langur is endemic to China and southwest Guangxi AR. It is found on karst formations covered with typical limestone forest. Caves and crevices are used as sleeping sites. It is distributed in isolated areas from the Shiwan Dashan Mountains north to the Zuo River, specifically in the karst hills of Nonggang Nature Reserve and Chongzuo Nature Reserve.

Appearance

The white-headed langur has a dark brownish to blackish body covered with long hairs. Its head is white to ivory col-

Urine stains

ored, with hairs forming a crest. The white color extends down to the shoulders, upper back, and upper chest. The tail is brownish at its base and white from the midpoint to the tip. The young have a bright orange coat with a naked pale face.

Scat

Hind footprint

Sleeping site

Habitat

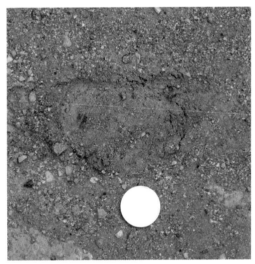

Track

Phayre's Leaf Monkey

Trachypithecus phayrei // Fei Shi Ye Hou

Adult female

IUCN: EN
CITES: II
CPS: I
CRL: VU

BL: 55–71 cm
TL: 60–80 cm
WT: 6–9 kg

are darker than the rest of the pelage. The lips and the area around the eyes are whitish. The pelage of the infants is orange brown until 3 months of age.

Distribution

The species occurs in Southeast Asia into southwest China. Two subspecies are recognized in China: *T. p. crepuscula* occurs in southern and southwest Yunnan Province, and *T. p. shanicus* occurs in western Yunnan Province, with the Salween River as the boundary between the subspecies. The species occurs in primary and secondary evergreen and semievergreen forest, in mixed moist deciduous forest, and also in bamboo-dominated areas, in degraded woodlands, and near tea plantations.

Appearance

The pelage of Phayre's leaf monkey is dark gray on the dorsal side and white on the ventral side. The head and tail

Habitat

Macaques
Mi Hou

General Description
Macaques (genus *Macaca*) are typically characterized by the presence of cheek pouches to carry food. All but one of the 23 macaque species are found in Asia, from Afghanistan to Japan and including the Philippines and Borneo. Although a short or absent tail characterizes most macaque species, some species have very long tails. The species have an intricate social structure, are highly adaptive, and display a great variety of calls and facial expressions.

Habitat and Ecology
The macaques occupy a wide geographic and ecologically diverse range. Although most monkeys are arboreal, macaques tend to be terrestrial or semiarboreal. The distribution of macaques is from sea level to over 4,000 m elevation and from semidesert shrub to moist temperate evergreen forests throughout central, South, and Southeast Asia.

Scat
Macaques produce brownish scats, and the scat size and shape will vary according to diet. The length of the pellets is generally 10–15 cm, with a diameter of 2–5 cm. Scats are oval shaped with a point at one end. The surface of the scats may have ruffles. Scats can be loosely formed and composed of seeds, leaves, and other vegetation.

Tracks
Macaques have wider and shorter feet compared with those of arboreal primates. The length of their footprint is only slightly longer than the width for both the hind and fore feet. The hind print has four fingers close together, with the thumb pointing in a perpendicular direction. The fore print is smaller than the hind print, with the four fingers more widespread.

Other Sign
Macaques feeding on tree seeds often leave partially consumed seeds and remnant shells on the ground under the trees.

Vocalizations are common among macaques. The most common vocalization is the "kaa" heard in a variety of contexts, but especially when the animal is being approached or threatened by humans or other animals.

Assam Macaque

Macaca assamensis // Xiong Hou

Subadult

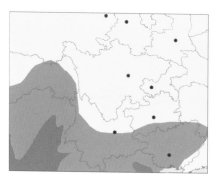

IUCN: NT
CITES: II
CPS: I
CRL: VU

BL: 52–67 cm
TL: 17–25 cm
WT: 5–12 kg

Distribution

The Assam macaque ranges through northern Southeast Asia and the Himalayas into southwest China. In the range of this guide it occurs in eastern Tibet AR, Yunnan Province and Guangxi AR, and southern Guizhou Province. It inhabits dense forest at the elevation range of 1,000–2,000 m and usually does not occur in secondary or degraded forest.

Appearance

The Assam macaque has a yellowish-gray to dark brown pelage. The facial skin is dark brownish to purplish. The head has a dark fringe of hair on the cheeks directed backward to the ears. The hair on the crown is parted in the middle. The shoulders, head, and arms tend to be paler than the hindquarters, which are grayish. The tail is well haired and short.

Scat

Habitat

Northern Pig-Tailed Macaque

Macaca leonine // Bei Tun Wei Hou

Adult male

IUCN: VU
CITES: II
CPS: I
CRL: CR

BL: 44–62 cm
TL: 12–18 cm
WT: 6–14 kg

Distribution

This species occurs in southwest Yunnan Province. It occupies elevations up to 2,000 m in tropical evergreen and semievergreen forest, tropical moist deciduous forest, coastal and swamp forest, low-elevation pine forests, and montane forest. It can occupy degraded forests.

Appearance

This macaque is characterized by its short, pig-like tail, which it normally carries in an erect backward arch over its back, with the tip partially resting on the rump. The macaque possesses a relatively long, uniformly golden-brown coat, with markings confined only to the brown crown, buff-colored cheek whiskers, and the red streak extending from the outer corner of each eye.

Rhesus Macaque

Macaca mulatta // Mi Hou

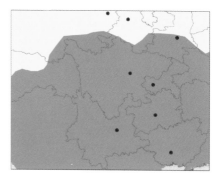

Female and subadults

IUCN: LC	BL: 43–60 cm
CITES: II	TL: 15–32 cm
CPS: II	WT: 5–10 kg
CRL: LC	

Distribution

Rhesus macaques have the widest geographic range of any nonhuman primate, occupying a great diversity of habitats, including grasslands, woodlands, and mountains of up to 3,500 m elevation. The species is distributed throughout central and southern China and can be abundant in areas with high human density. It is found within all regions covered by this guide with the exception of the highest and most northern plateau areas.

Juvenille

Appearance

Rhesus macaques are medium sized, and the upper back varies from yellowish gray to golden brown, whereas the lower back is usually more brightly colored. The pelage on the abdomen is pale buff or gray and sparser than the dorsal side. Thinly haired facial skin is pale brown to reddish. The tail is of medium length. Adult males are larger and heavier than females.

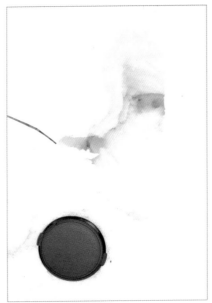

Tracks

Stump-Tailed Macaque

Macaca arctoides // Duan Wei Hou

Adult

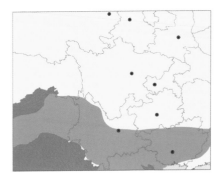

IUCN: VU
CITES: II
CPS: II
CRL: VU

BL: 48–65 cm
TL: 3–7 cm
WT: 7–10 kg

Distribution

This species is found across northern Southeast Asia and parts of the Himalayas into southwest China. In the region of this guide it is found in Yunnan Province and Guangxi AR (south of 25°N). It ranges from tropical semievergreen forest to tropical wet evergreen and deciduous forest at elevations up to 2,500 m.

Appearance

Stump-tailed macaques have thick, long, dark brown fur covering the body, as well as the short tail. They have bright pink or red faces that darken to brown or nearly black as they age and are exposed to sunlight. Adult males are much larger than females and have elongated canine teeth which are absent in females.

Tibetan Macaque
Macaca thibetana // Zang Qiu Hou

Female and juveniles

IUCN: NT
CITES: II
CPS: II
CRL: VU

BL: 51–70 cm
TL: 4–14 cm
WT: 9–25 kg

Distribution

This species is found in east central China. In the range of this guide it is located in Guizhou and Sichuan Provinces, northern Guangxi AR, Yunnan Province, and Tibet AR at elevations from 800 to 3,000 m. The species occurs in tropical or subtropical mixed deciduous to evergreen forest. Two large populations are distributed in Emei and Huangshan Mountains in the tourism region, often begging food from tourists.

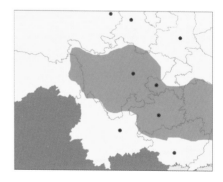

Appearance

The Tibetan macaque is the largest species of macaque and one of the largest monkeys in Asia. The fur is brown on the

Scat

Partly eaten kiwi fruits

dorsal side and creamy buff to gray on the ventral side, with a prominent, pale-buff beard and full-cheek whiskers framing the hairless face. The bare facial skin is pale pink in males and a more vivid, reddish pink in females, particularly around the eyes. A long, dense coat helps this species cope with the cold environment at high altitudes. The species has a short, stump-like tail. Adult males are larger and heavier than females.

White-Cheeked Macaque
Macaca leucogenys // Bai Jia Mi Hou

In its habitat

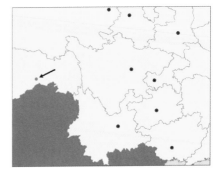

IUCN: NE
CITES: NL
CPS: NL
CRL: CR

BL: Not yet measured
TL: Not yet measured
WT: Not yet measured

Distribution

The white-cheeked macaque was first described in 2015, with very limited infor-

Adult

mation about its distribution. The range shown is based on local knowledge since this species has not yet been assessed by the IUCN. The species was observed in tropical forest at an altitude of 1,395 m and in primary and secondary evergreen broadleaf forest up to 2,700 m in Mêdog County, Tibet AR.

Appearance

The white-cheeked macaque and Assam macaque were believed to be the same species until 2015. The most noticeable difference between the sympatric species is the presence of white, elongated whiskers on the face of the white-cheeked macaque. The whiskers begin to grow as individuals the creatures approach sexual maturity and eventually cover the whole face, giving the animals a rounded facial appearance. The white-cheeked macaque can be distinguished from other macaque species by a suite of characteristics, including relatively uniform brownish dorsal pelage, hairy and lighter-colored dorsal pelage, relatively hairless short tail, dense hair along the neck, prominent pale to white side and chin whiskers, and dark facial skin.

Slow Loris
Feng Hou

General Description
Slow lorises (genus *Nycticebus*) are found in Southeast Asia and bordering countries, including parts of Bangladesh and northeast India in the west and the Philippines' Sulu Archipelago and the island of Java in the southeast. Within the range of this guide, they are found primarily in Yunnan Province. Slow lorises have a round head, narrow snout, large eyes, and species-specific coat patterns. They are arboreal and nocturnal; they sleep during the day, rolled up in a ball above the ground, often hidden in trees palm fronds or lianas. They start their nightly activity around sunset and exhibit relatively slow movements compared with other mammals.

Habitat and Ecology
Slow lorises are generally found high in the trees in tropical rainforests, preferring forests in warm, lowland areas below either 2,400 m (Bengal slow loris) or 1,500 m (pygmy slow loris) in elevation. The slow loris species in China are perhaps the most edge adapted of the genus, staying near the rainforest boundaries where vertical supports are abundant.

Scat
Small fecal pellets (0.5-cm diameter) are rarely found as they scatter when dropped from arboreal sites.

Tracks
All lorises have extremely strong fingers and toes and are capable of maintaining a powerful grip on branches for long periods of time. The species come to the ground under only rare conditions, such as the need to move between isolated trees. No tracks have been reported in the wild.

Other Sign
As with many nocturnal species, the tapetum at the back of their eyes reflects light, giving off a bright orange-red eyeshine at night. Loris eyeshine is most commonly observed by scanning vegetation with a spotlight. We do not know if the species differ in the color or shape of their eyeshine. Slow lorises produce several sounds, ranging from a low hiss or growl when disturbed to a high whistle made by females in estrus.

Bengal Slow Loris
Nycticebus bengalensis // Feng Hou

Adult

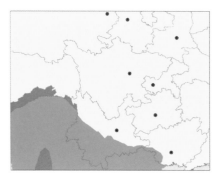

IUCN: VU
CITES: I
CPS: I
CRL: EN

BL: 26–38 cm
TL: 22–25 cm
WT: 1–2 kg

Distribution

The Bengal slow loris is sympatric with the pygmy slow loris in southeast of China, Vietnam, and Laos. In China, it is found in southern Yunnan Province and possibly southern Guangxi AR.

Appearance

The Bengal slow loris is the largest species of slow loris. It has dense, woolly, brown-gray fur on its back and white fur on its underside. It also has a clear dark stripe that runs up to the top of its head but does not extend laterally toward the ears. Its forearm and hand are almost white. Like other slow lorises, its tail is vestigial, and it has a round head and short ears. It has a rhinarium (the moist, naked surface around the nostrils of the nose) and a broad, flat face with large eyes. Its eyes reflect a bright orange eyeshine at night when a light is directed toward them.

Pygmy Slow Loris

Nycticebus pygmaeus // Wo Feng Hou

Adult

IUCN: VU
CITES: I
CPS: I
CRL: CR

BL: 21–26 cm
TL: 10 cm
WT: 0.3–0.8 kg

Distribution

The pygmy slow loris is found east of the Mekong River in Vietnam, Laos, eastern Cambodia, and China. In China it has been recorded only from Pingbian, Hekou, Jinping, and Luchun Counties of Yunnan Province.

Appearance

The pygmy slow loris is a small, compact primate. It has a short tail, a short and rounded muzzle, large, round eyes directed forward, and short, dense fur. It is mainly reddish brown or gray with a white line between its eyes and dark markings around its eyes. Hands are broad with an opposable thumb. Males and females are similar in appearance.

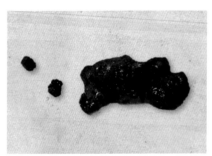

Scat

Snub-Nosed Monkeys
Jin Si Hou

General Description
Snub-nosed monkeys are primates of the genus *Rhinopithecus*, whose name derives from the short, upturned nose on their round face. They have relatively multicolored and long fur, particularly at the shoulders and back. The genus occurs in low densities across its range. They are endemic in Asia, with a range covering southern China (especially Tibet AR, Hubei, Shaanxi, Sichuan, Yunnan, and Guizhou Provinces) as well as the northern parts of Vietnam and Myanmar.

Habitat and Ecology
Rhinopithecus species are among the few primates who live in temperate zones. Snub-nosed monkeys inhabit mountain forests up to a height of 4,800 m. They spend the majority of their life in the trees but may forage on the ground and migrate to lower elevations during the winter. Snub-nosed monkeys live together in large groups but split up into smaller groups in times of food scarcity.

Scat
Scats are a series of individual pellets (1 × 3 cm) connected like a string of prayer beads, but discrete pellets are more often found after the pellet string falls from a tree. Fresh scats are soft and greenish, whereas the old ones are dry and brownish. Scats contain lichens, seeds, pine needles, and insects, depending on the diet and season.

Tracks
Tracks can occasionally be found on snow or a muddy surface where the animal foraged on the ground. The footprints (10–20 cm in length, 4–6 cm in width) are slender and have five toe marks.

Other Sign
Snub-nosed monkeys typically feed on tender bark. They break branches before feeding and leave the barkless branches both in the tree and scattered on the ground. Group movement may also cause broken branches under their travel route. They sometimes forage on the ground and turn over moss to search for insects. As social animals, they live together in large groups of up to several hundred members. They have a large vocal repertoire, from solo calls to a group chorus. The most common vocalization heard by humans is the "kaa" or "woo-kaa" in a variety of contexts, but this call is common as an alarm when they are approached by humans.

Black Snub-Nosed Monkey

Rhinopithecus bieti /// Dian Jin Si Hou

Adult male

IUCN: EN
CITES: I
CPS: I
CRL: EN

BL: 74–83 cm
TL: 51–72 cm
WT: 12–17 kg

Distribution and Habitat

This species occurs in southwest China, southeast Tibet AR, and northwest Yunnan Province, living in temperate coniferous forest or deciduous/evergreen broadleaf and coniferous forest at an altitude of up to 4,700 m.

Appearance

The black snub-nosed monkey is black on the dorsal side and white below, with a gray face and a forward-curling tuft of hair on the crown of the head. The periphery of the eyes and the area around the mouth and nose are steel gray, and the lips are flesh colored. The coat color is distinct from that of other snub-nosed monkeys, with a black torso, and the outer surfaces of the limbs and tail are brownish gray or dark gray. The shoulders of the male have long dark gray hair. The color patterns of males and females are similar, but the female is comparatively smaller than the male, and its hair is noticeably shorter.

Scat

Habitat

Golden Snub-Nosed Monkey

Rhinopithecus roxellana // Chuan Jin Si Hou

Adult male

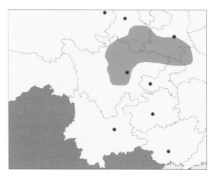

IUCN: EN
CITES: I
CPS: I
CRL: VU

BL: 52–78 cm
TL: 57–80 cm
WT: 9–25 kg

Distribution

The golden snub-nosed monkey inhabits temperate forests on mountains in four provinces in China: Sichuan, Gansu, Shaanxi, and Hubei. Within the range of this guide it occurs in western Sichuan across five mountain ranges (Qionglai, Minshan, Daxiangling, Xiaoxiangling, and Daba Mountains) and also in southern Gansu (Minshan Mountains) and southern Shaanxi (Qinling Mountains).

Habitat and Ecology

This species is found at elevations of 1,500–3,400 m. Over this range the habitat varies from deciduous broadleaf forests at lower elevations to mixed coniferous broadleaf forests and coniferous forests above 2,000 m.

Appearance

The species has long and dense hairs, generally yellowish red and overlaid with black on the back. The ventral side is medium brown, whereas the dorsal side, crown to nape, arms, and outer thighs are deep brown. Males are similar to females but larger and with brighter tones. Newborns (estimated at less than 3 months of age) are dark to light gray in coloration.

Adult (front) and subadult (back)

Branch broken by golden snub-nosed monkey

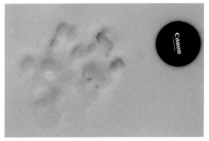

Track

Scat

Feeding site

Gray Snub-Nosed Monkey

Rhinopithecus brelichi // Qian Jin Si Hou

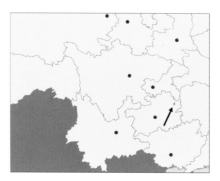

Adult female (top) and male (bottom)

IUCN: EN
CITES: I
CPS: I
CRL: CR

BL: 64–69 cm
TL: 70–85 cm
WT: 8–15 kg

Distribution and Habitat

This species is endemic to a small region of Guizhou Province in southern China (in Jiangkou, Songtao, and Yingjiang Counties). It is confined to a small, continuous block of habitat centering on Fanjing Mountain, south of the Yangtze in the Wuling Mountains. The species lives at elevations of 1,400–2,300 m in summers and moves down to 570 m elevation at times of heavy snow cover. It inhabits mixed deciduous and evergreen broadleaf forest, including secondary forest. There were unconfirmed reports of a population in Jinfoshan Nature Reserve, Chongqing City.

Appearance

The gray snub-nosed monkey is covered with long, fine hair, generally grading from brown on its upper body to gray on its lower body, with a white patch between its shoulder blades. Its head, its neck, and the ends of its limbs are black, except for a golden brow. The monkey has a golden chest and chestnut color on its inner knees and the inner sides of its upper arms. Its face is bare bluish-white skin, with pink skin around the eyes and mouth. Adult males are brighter colored and longer haired than adult females and have visible white skin around their nipples.

Subadult

Myanmar Snub-Nosed Monkey

Rhinopithecus strykeri // Mian Dian Jin Si Hou

Adult male

IUCN: CR	BL: 55 cm
CITES: I	TL: 78 cm
CPS: I	WT: 12-20 kg
CRL: CR	

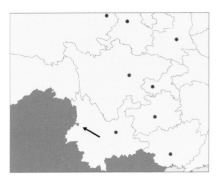

Distribution and Habitat

In 2011, populations were discovered in Lushui County and in Gaoligongshan National Nature Reserve in Yunnan Province. The range shown is increased from that indicated by IUCN on the basis of our knowledge of these recent sightings. The species spends its summer months in temperate mixed forests at higher elevations (up to 3,200 m) and descends to lower elevations (down to 1,700 m) in the winter to escape snow. We have no information about its habitat requirements.

Appearance

The fur of the Myanmar snub-nosed monkey is mostly black. Its crown consists of a thin, high, forward-curved crest of long, black hairs. It has protruding white tufts of hair around its ears, a mostly naked face with pale pink skin, a "moustache" of whitish hairs above the upper lip, and a distinct white chin beard. The limbs are mostly black; the inner sides of the upper arms and upper legs are blackish brown.

Adult male (left) and female (right)

Dhole
Cuon alpinus // Chai
Asiatic Wild Dog, Indian Wild Dog, Red Dog

Adult

IUCN: EN
CITES: II
CPS: II
CRL: EN

BL: 88–113 cm
TL: 40–50 cm
WT: Male, 15–20 kg;
     Female, 10–17 kg

Distribution

The dhole is widely, but patchily, distributed in central, East, South, and Southeast Asia and Russia. It was historically reported in most of China, but there are few records in recent decades. Its current distribution is poorly known but must be highly fragmented. Confirmed records within the region are fewer than 10 sites in southern and western Gansu Province, southern Shaanxi Province, western Sichuan Province, western Yunnan Province, and southeast Tibet AR.

Appearance

The dhole is a medium-sized canid with a short muzzle and tail. Its coat color is grayish red to dark red on the back and body sides, with a paler underside. It has white fur around the mouth and chin. The ears are big and round, with white inside. The tail is bushy and black.

Habitat and Ecology

Dholes are found in multiple habitats across their range from open meadows to dense forests and semiarid desert. Dholes live in packs of up to 12 or more individuals that cooperate in hunting large ungulates. Prey includes wild boars, deer, and wild sheep and goats, as well as smaller animals like rodents and hares. Dholes will also scavenge on animal carcass. Dholes generally breed in spring and give birth to four to six pups per litter, and the pups are raised cooperatively. During the past three decades, the wild populations of dholes in southwest China have suffered severe declines and reduced range. Use of poisons as retaliatory killings after dholes prey on livestock are speculated to be the most probable cause.

Scat

Scat

The shape of the dhole scat varies depending on diet, with scats composed of hair and bone shaped like tapered ropes. Their scats can be dilute when the diet is mostly meat or internal organs. When other sympatric carnivore species of similar size (e.g., Asiatic golden cat) are present, the scats of dholes may be indistinguishable.

Tracks

The dhole's track is a midsized print (5 × 7 cm). It is symmetric, with four of the five front toes registered on the track, and the first toe is higher than the others. The plantar pad is triangulate and proportionally small. The claws are all visible.

Track

Raccoon Dog

Nyctereutes procyonoides // He

Adult

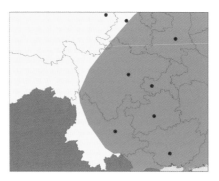

IUCN: LC
CITES: NL
CPS: NL
CRL: NT

BL: 45–70 cm
TL: 16–22 cm
WT: 3–12.5 kg

and tail is gray with a black tip, and hair on the legs and feet is dark brown. The tail length is less than one-third of the body length.

Distribution

Raccoon dogs are distributed in East and Northeast Asia and have been introduced into Europe. They have been widely farmed as fur bearers; therefore, many farm escapes have established local populations both in China and abroad. In our focus area, they are reported, but not common, in all provinces east of the plateau region, Guangxi AR, and Chongqing City.

Appearance

Raccoon dogs appear more like a procyonid rather than a typical canid, primarily because of their stocky body, proportionally short legs, short tail, small and round ears, short muzzle, and black facial mask. The face appears masked with a whitish or light gray forehead and muzzle, with black around the eyes. Long fur on the cheeks forms a ruff on the neck. The hair on the body

Habitat and Ecology

Raccoon dogs are found in open areas, such as open broadleaf forest, meadow, brushland, and marshland, and often near water. Raccoon dogs prefer to forage in open woodlands with rich

Scat

understories dense with ferns. They are more omnivorous than most other canids, although their primary prey is rodents. They will feed on rodents, amphibians, mollusks, fishes, insects, birds and their eggs, and the plant parts of roots, stems, seeds, fruits, and nuts. They are nocturnal and normally solitary but are sometimes observed in pairs or family groups. They are thought to be monogamous and breed in the spring, with a litter size of five to eight young. Although their range is widespread, they have not been encountered by the authors and must be rare within the range.

Tracks

We have no details of tracks in the region.

Red Fox
Vulpes vulpes // Chi Hu

Adult with summer coat

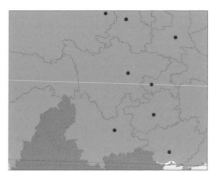

IUCN: LC
CITES: III
CPS: NL
CRL: NT

BL: 50–90 cm
TL: 35–45 cm
WT: Male, 4–14 kg;
    Female, 3.6–7.5 kg

Distribution

The red fox is the most widely distributed carnivore in the world, and it occurs in most of the Northern Hemisphere. The species is historically found throughout our focus area, but its current distribution is poorly known. Recent confirmed records are from western and southern Sichuan, northwest Yunnan, western and southern Gansu, and Qinghai Provinces and the Tibet AR.

Appearance

The red fox is a small-sized canid, with relatively long and thin legs. Adult males are larger than females. The coat color can vary from yellow to brown or dark red. In this region of China, the red fox's coat color is reddish brown on the back, yellow on the shoulder and body sides, and white on the underside. The tail length is longer than half of the body length, and the tail is bushy and the same color as the body with a white tip. The winter coat is much thicker and paler than the summer coat. Compared with the sympatric Tibetan fox (on the Qinghai-Tibetan Plateau), the latter has longer legs, larger ears, and a longer muzzle. Another distinct characteristic of the red fox is that the back of the ears is black or dark brown.

Habitat and Ecology

Red foxes are found in all habitats, including forest, shrubland, meadows, agriculture, semideserts, and high-elevation tundra, and even among human residences. They can occur up to 4,500 m elevation. Their diet is omnivorous and includes small rodents, rabbits, pikas, birds, amphibians, reptiles, insects, fruits, and vegetables. Red foxes rely extensively on carrion in winter and early spring. They are typically solitary, although small mother-young groups are also frequently observed. Red foxes are monogamous, and the males are involved with parental care. Females give birth from March to May, and litter size varies from 1 to 10 kits.

Scat

Red foxes deposit small to medium-sized carnivore scat (diameter typically <1.5 cm). Their scats are usually found on elevated surfaces (e.g., logs and rocks) along the trail or road or other open spots. The scats are composed of hair and bones and are twisted and tapered ropes. Compared with the scats of sympatric small felid species (e.g., leopard cat, Chinese mountain cat), red fox scats usually have no or a very short (<1 cm) needle-shaped tail at the end of the last segment.

Tracks

The red fox has small to medium-sized footprints (3.4–4.5 × 5.0–6.0 cm). The form of their prints is similar to those of domestic dogs, with separate imprints of four toes. The claw marks of all four toes are visible.

Other Sign

Red foxes create burrows for dens and shelter. Newly dug burrows can be visible at a distance with disturbed soil and small rocks. Red foxes frequently use old burrows left by other animals (e.g., marmots) as temporary shelters. Shed hairs of red foxes can be found at the burrow entrance, and their scats are also frequently found both inside and outside the entrance.

Adult with winter coat

Fresh scat

Burrow

Himalayan marmot burrow that was used by a red fox

Tibetan Fox
Vulpes ferrilata // Zang Hu

Adult with summer coat

IUCN: LC
CITES: NL
CPS: NL
CRL: NT

BL: 49–65 cm
TL: 25–30 cm
WT: 3.8–4.6 kg

Distribution

The Tibetan fox is endemic to the Qinghai-Tibet Plateau. In our focus area, it occurs in Qinghai and western Gansu and Sichuan Provinces and Tibet AR.

Appearance

The Tibetan fox has a brown-red coat color on the back with a white underside and broad light gray bands on its flanks. The reddish color is also found on the muzzle, head, neck, and legs. The small ears are light brown on the back and white inside. The tail is shorter than half of the body length, with a gray color and white tip. The winter coat is much thicker than the summer coat. Compared with the sympatric red fox, the Tibetan fox has a much broader face (the shape of the head is nearly square when observed from front), smaller ears, and a stockier body with shorter legs.

Habitat and Ecology

The Tibetan fox is found throughout the alpine region, including alpine meadows, semiarid and arid tundra, and steppe up to more than 5,000 m. The Tibetan fox has a body size and feeding niche similar to those of the red fox, with some overlap in their distributions. The most important prey of Tibetan foxes are small rodents and pikas, but they also can feed on lizards, hares, marmots, birds, and carrion. They are diurnal and solitary, although they are occasionally observed in pairs or with young. Their burrows are found in the base of a slope or along an old river bank. Tibetan foxes mate in late February, and females give birth to two to five kits per litter.

Scat

The Tibetan fox leaves a small- to medium-sized carnivore scat (typically, diameter of <1.5 cm). Its scat is similar to that of the red fox and cannot be differentiated in the field where the species are sympatric. The shape and color may vary broadly depending on region and diets. The scats are composed of hair and bones and are twisted and tapered ropes.

Tracks

The track of the Tibetan fox is similar to that of the red fox but smaller in size (3–4 × 4.5–5 cm). The Tibetan fox has a small to medium-sized print with four front toes and claw marks visible in the track. The plantar pad is proportionally small. When the two fox species are sympatric, their tracks may be indistinguishable.

Scat

Tracks

Wolf

Canis lupus // Lang

Gray Wolf, Grey Wolf

Adult with winter coat

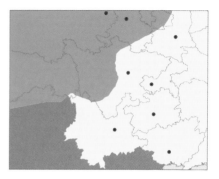

IUCN: LC
CITES: II
CPS: NL
CRL: NT

BL: 100–160 cm
TL: 38–55 cm
WT: 28–60 kg

Distribution

The wolf is a widely distributed species in the Northern Hemisphere. Within our focus area, wolves historically occurred across most of the northern provinces, but the current range is reduced to the remote alpine or arid/semiarid habitats in western Gansu and Sichuan Provinces, Qinghai Province, and the Tibet AR. The range shown is reduced from that indicated by IUCN on the basis of our local knowledge.

Appearance

The wolf is the largest canid species in the world. The typical pelage color is gray, although it varies broadly from grizzly, yellowish to brownish gray and dark gray. The winter coat is much thicker, denser, and darker than the summer coat. Compared with other canids,

the wolf has a longer muzzle. The eyes and ears are front pointing. The wolf looks thin for a large canid with relative long legs. The tail is bushy with no color changes along most of its length.

Habitat and Ecology

Wolves can use all natural habitats such as forest, shrubland, meadow, and tundra across all elevations, although re-

Adult with summer coat

cent reports of wolves within our focus region are mostly in remote areas of alpine or arid/semiarid habitats of the Qinghai-Tibetan Plateau and not in lower-elevation forests. Wolves feed primarily on large ungulates such as deer, blue sheep, gazelles, and wild boars, and they can also prey on smaller animals like marmots, hares, and birds. Wolves have good endurance and are capable of running and chasing large prey over a long distance (>10 km). They may also occasionally scavenge on animal carcasses and the remains of other sympatric carnivores such as snow leopards, leopards, and brown bears. There are increasing reports of wolves killing domestic livestock on the plateau. Wolves are social animals that live and hunt in family groups, with each individual having a distinct rank in the pack. The normal reported litter size is about six pups, and both parents raise the offspring.

Scat

Wolf scat can be as much as 4 cm in diameter. The shape varies depending on

Fresh scat

Old scat

Track

Track

the food consumed, with hair and bone most often found; scats are usually twisted and tapered ropes with needle ends. Scats composed of meat and internal organs could be amorphous or tubes with blunt ends and rapidly degrade. Fresh scats are normally gray to black depending on diet and have strong odor. Old scats are light gray to white and may remain in the field over several months in arid or cold environments.

Tracks

The wolf track is large (6–7 × 4.5–5.5 cm) and symmetric, with four of the five toes registered on the track, and the first toe is higher than the others. The plantar pad is triangulate. The most dis-tinctive characteristic of a wolf track, compared with a felid track, is the claw marks, as all nails register clearly for the wolf. The metacarpal pad is also proportionally smaller when compared with a felid track.

Other Sign

Howling is the special behavior of wolves and is used to declare their territories. Loud barks may also be occasionally heard over long distances when they hunt. Females give birth and raise the offspring in dens that they dug themselves or were left by other animals or sometimes in natural caves and rock crevices. Numerous shed hairs are usually found around the entrance of the den.

BEARS // XIONG

Asiatic Black Bear
Ursus thibetanus // Hei Xiong
Moon Bear, Himalayan Black Bear

Side view of adult

IUCN: VU
CITES: I
CPS: II
CRL: VU

BL: 116–175 cm
TL: 5–16 cm
WT: 54–240 kg

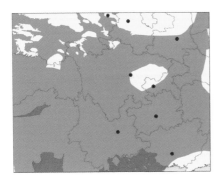

Distribution

Asiatic black bears were historically distributed across most of temperate, subtropical, and tropical Asia (including large islands near the mainland) but are now patchily distributed in East Asia, Southeast Asia, and South Asia. Within the region of this guide, the Asiatic black bear is found in all provinces and ARs.

Appearance

The Asiatic black bear is a medium-sized, black-colored ursid, with a gray to dark brownish muzzle and round ears. Its body is stocky, and its short limbs are

Juvenile

strong with broad paws and long claws. The short tail is not conspicuous in relation to its body length. The most distinctive characteristics of the species are the V-shaped white patch under the chin and a crescent patch on its chest that accounts for it sometimes being called a "moon bear." The chest patch is variable in size and shape, allowing for individual identification of animals. Adults may have a mane of black hairs along the neck that is rather long and thick and makes the neck appear thick.

Habitat and Ecology

Asiatic black bears occupy forested habitats, both broadleaf and coniferous, from near sea level to 4,000 m elevation. They may infrequently use open alpine meadows at the highest elevations. They are omnivores, and their diet depends on season and availability; diets include succulent vegetation in spring, insects and tree or shrub-borne fruits in summer, and nuts in autumn. The autumn mast production plays an important role in their annual nutrition intake, in part to put on sufficient fat reserves for winter denning. The hives of wild and domestic bees are destroyed for the honey and combs. Asiatic black bears also scavenge carcasses of large mammals when possible. They occasionally predate on livestock and frequently forage in croplands, which leads to extensive crop damages and therefore causes severe human-bear conflicts. In cold areas of this region (northern areas and elevations higher than 2,500 m), where food becomes scarce in winter, both sexes hibernate in rock crevices, on ledges, or in tree dens. Bears enter dens as early as October and exit as late as May, but adult male bears can be active throughout the winter. Asiatic black bears are solitary animals and breed during June–July. Females give birth during November–March in the den. Age of first reproduction is 4–5 years, and females normally produce litters of one to three cubs every other year. Mother-offspring groups of two to four individuals are also frequently observed.

Asiatic black bears are poached for meat, and their body parts are used in traditional Chinese medicine (e.g., the gallbladder and fat). Retaliatory killings and poisoning are also common in places with frequent human-bear conflicts.

Scat

The form, shape, and color of Asiatic black bear scats vary broadly depending on their diets. Scats with animal re-

Scat (berry diet)

Fresh scat (animal and plant diet)

Fresh scat (acorn mast diet)

mains are usually found in piles, with a needle point at the end of one segment. Normally, scats are more dilute in the early spring because of the diet of new leaves and forbs. It is common to find the leftover components of fruit in summer and mast in autumn (berry seeds, mast shells, etc.).

Tracks

Asiatic black bear tracks are large (13–19 cm). The track shows five toes and claws. The print of the forefoot is wider than that of the hind foot. The hind foot track is more deeply imprinted on the outer side because of its pigeon-toed walking style. The track of the Asiatic black bear can be differentiated from that of the sympatric giant panda in snow or mud since foot hairs are visible on the fore print of the Asiatic black bear but not on that of the giant panda.

Hind foot track

Other Sign

Asiatic black bears leave a large amount of sign that is more detectable than scats or tracks. When Asiatic black bears feed on mast seeds (acorns, walnuts, chestnuts, etc.) in mature trees, they often break branches toward themselves while sitting in a fork of major limbs or along the trunk. These broken branches

Forefoot track

accumulate in these junctures and form a feeding platform, or "nest." This is distinctive sign of Asiatic black bears in this region and can be observed from a distance on steep slopes.

Asiatic black bears are good tree climbers and leave distinctive claw marks on the trunk when climbing. Normally, there are three or four parallel scratches 2 to 3 cm apart from each other. The scratches are oblique to the trunk for the fore claws and parallel to the trunk for the hind claws. Claw marks left on trees with coarse bark (e.g., alpine oaks) may appear as only pointed scratches. Giant pandas will also climb trees, and it is hard to differentiate

marks, but giant pandas do not make tree feeding platforms, and the feeding sites of giant pandas contain copious fecal droppings.

Asiatic black bears will feed on the sap and soft wood under the bark of coniferous trees in spring. This feeding involves removing the bark across a large patch of a mature tree and using their bottom incisors to scrape the sap. A series of parallel grooves is left on the inner tree. The debarked patches have been observed only on lower sections of trees (1–2 m).

Asiatic black bears feed on ants and their eggs by turning over rocks and logs and searching underneath for ant nests, activities that leave unique sign such as disturbed rocks and logs. Such signs are frequently found in dry forests with abundant ant colonies, such as in western Sichuan and Yunnan Provinces.

Feeding sign on oak tree

Ant nest under rock disturbed by Asiatic black bear

Claw marks on coarse bark

Claw marks on smooth bark

Feeding sign on coniferous tree

Feeding platform

Brown Bear

Ursus arctos // Zong Xiong

Horse Bear, Tibetan Brown Bear

Adult brown bear scavenging

IUCN: LC
CITES: I
CPS: II
CRL: VU

BL: 115–119 cm
TL: 8–13 cm
WT: 125–225 kg

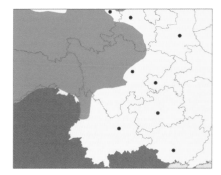

Distribution

The brown bear is the most widely distributed ursid in the world. In the region of this guide, the brown bear is present only at high elevations above or near the tree line on the Qinghai-Tibetan Plateau, including western Sichuan, Yunnan, and Gansu Provinces, as well as Qinghai Province and the Tibet AR.

Appearance

Compared with the brown bear in other regions of the world, the brown bear on the Qinghai-Tibetan Plateau is relatively small bodied, but it is still the largest carnivore in this region. The bear's coat color in this region varies from black, dark brown, reddish brown, or light brown to gray and even completely white. Regardless of the dominant coat color, the color is usually variegated, with the four limbs darker and the body and head lighter in color. Many individuals have a whitish or light yellowish patch around the neck that may extend to the shoulder and chest, although the patch size varies broadly and may be totally absent in some individuals. The massive musculature in the shoulders gives the bear a humped shoulder.

Habitat and Ecology

Worldwide, brown bears occupy most terrestrial habitats from forest to grasslands and even deserts. In the Qinghai-Tibetan Plateau area they are found at elevations of up to 5,000 m, and brown bears use the plateau grassland, shrubland, and the edge of coniferous forests. The bear's diet depends on seasonal prey availability, with a large proportion composed of vegetation, including grasses, forbs, bulbs, roots, and tubers. The routine prey of brown bear are pikas and marmots. They also rob carcasses of ungulates and livestock from other predators such as snow leopards and wolves. Currently, the brown bears on the plateau have increasing conflicts with the local people because they break into houses for human food and occasionally prey on livestock. The bears normally go into hibernation in late October and emerge in early May, but recent observations show the delay of hibernation, possibly because of warming climate or poor body condition related to food shortages. Brown bears generally breed during early May–July, but females delay implantation until about October or November and give birth during the hibernation. Age of first reproduction is 4.5–7 years, and the average litter size is two cubs produced in alternate years.

Scat

Brown bear scats are typically found in extremely large piles. The size (diameter of 4–6 cm) is the largest among all carnivore species within the region. The scat usually contains undigested food that is readily identifiable, such as hairs, bones, skin, and plant fibers. Fresh scats have a strong odor, and the color varies from black, brown, or yellow to gray depending on diet.

Flat scat

Segmented scat

Tracks

The large track shows five toes and claws, and the fifth toe (the outermost one) is the biggest. The fore print (14 × 17 cm) shows a smaller footpad than the hind foot (17 × 33 cm). In snow, the hind track can look like a big human footprint.

Other Sign

Brown bears leave obvious evidence of digging on the grassland when they try to extract pikas or marmots from their burrows. The shape and depth of the digging depends on the depth of the burrow and their speed of capture, but extensive diggings are indicative of brown bears.

Brown bears normally bed or den under rock cliffs or in caves. These bedding and den sites normally appear as a large, oval-shaped depression on the ground, usually found with numerous shed hairs in the pit and sometimes scats nearby.

Adult

Hind foot track

Forefoot track

Bedding site

Tracks

Giant Panda
Ailuropoda melanoleuca // Da Xiong Mao
Panda

Adult

IUCN: VU
CITES: I
CPS: I
CRL: VU

BL: 120-180 cm
TL: 12-20 cm
WT: 75-125 kg

Distribution
Giant pandas have a highly fragmented distribution restricted in six mountain ranges (Qinling, Minshan, Qionglai,
Daxiangling, Xiaoxiangling, and Liangshan Mountains) across three provinces

in southwest China: Shaanxi, Gansu, and Sichuan. In 2015 there was a record of an individual in northeast Yunnan Province to the south of the Yangtze River.

Appearance

The giant panda is a large, solid-built bear, with a distinct black-and-white coat, large round head, and relatively short and blunt muzzle compared with that of other bears. Adult males are larger than females. Its limbs, shoulder, and ears are black, and the rest of the body is mostly white. A rare brown-and-white form, on which the black is replaced by light brown or cinnamon, has been reported in Qinling Mountains, Shaanxi Province. The coat color pattern of young and juvenile bears is similar to that of adults, whereas the body shape is more rounded.

Habitat and Ecology

Giant pandas occupy temperate forest with dense stands of bamboo at elevations of 1,200–3,200 m. They have a specialized diet of almost 100% bamboo, although they feed on numerous species of bamboo. They also occasionally scavenge on animal carcasses in the wild. They can conduct annual elevational migrations following the annual growth cycle of bamboo. An individual's home range is 4–29 km2 (mean of 10 km2) and may vary depending on food availability. Giant pandas occasionally climb trees during mating to rest or avoid threats, especially young individuals. Giant pandas are solitary animals, except for the mother-young pair. Adults mate in spring (March–May), and the males often fight for access to estrous females. Females reproduce once every 2–3 years and give birth during mid- to late summer (August–early October) to a single cub or, occasionally, twins in hollow trees, rock crevices, or caves. Juveniles reach sexual maturity at approximately 4.5 years old. Unlike other bears, giant pandas do not hibernate during winter. They normally spend much of the day feeding, up to 12–14 hours per day, and the remainder is spent sleeping.

Scat

Adult giant pandas will consume 10–15 kg of poorly digestible bamboo daily; therefore, they produce a large amount of feces. Their scats are unique in size and shape and cannot be confused with those of other animals. The large (13–21 × 4–8 cm), oval-shaped scat contains fairly uniform sized fragments of bamboo stems and/or bamboo leaves. Fresh scats are greenish, and old ones are yel-

During mating season

Adult with rare brown-and-white coat

Fresh scat

Dry scat

Large scat deposits at feeding site

Rare scat containing animal remains

lowish and loose. Scat may be found in large piles near feeding or denning sites. Scats of the red panda also contain bamboo remains with a similar shape but are much smaller in size (3–4 cm in length).

Giant pandas occasionally prey on livestock or scavenge on wildlife carcasses and therefore produce rare scats containing hair, bone chips, and other undigested animal remains. These scats have regular size and shape but are normally covered by a thick white coating and have a strong putrid odor.

Tracks

Giant panda tracks (15–18 × 9–12 cm) can be found on muddy substrate or snow but are hard to distinguish from those of the Asiatic black bear. Tracks of the panda's front paws obviously turn inward, that is, are pigeon-toed, compared with those of the Asiatic black bear.

Other Sign

Giant pandas spend 12–14 hours daily feeding and leave obvious feeding sign, frequently accompanied by scats, at the feeding sites. This feeding sign normally includes debris from bamboo stems or shoots and shoot shells left in small piles on the ground.

Giant pandas are solitary and typically communicate through scent marking by spraying urine and rubbing a secretion from their anal glands onto tree trunks. Scent marks can be visually found on trees or logs, especially during the mating season (March–early May). The scent trees chosen by pandas vary broadly in size (diameter at breast height ranges from 2 to 100+ cm) and species (both conifer and broadleaf trees). Fresh scent marks have a slightly sour and musky smell and are frequently found with hairs and fresh bite marks. Old scent marks may look just like a darker stain on the tree bark and cannot be easily recognized.

Female giant pandas give birth to their young in well-secluded dens during summer. In the Qinling Mountains, giant pandas mainly select rock caves and crevices as den sites, whereas in other mountain ranges they primarily use near-ground hollows of big trees (mostly conifers, occasionally broadleaf trees). Large amounts of scat are usually found near the den sites, and shed hairs can be found in the den or on the rock and tree bark at the den entrance.

Track

Remains after feeding

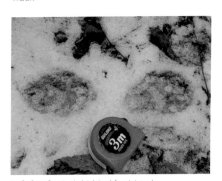

Left forefoot, right hind foot track

Fresh feeding sign and scats

Tree hollow used as birthing/nursing den

Rock cave used as birthing den

Scent mark

PROCYONID

Red Panda

Ailurus fulgens // Xiao Xiong Mao
Red Cat-Bear, Lesser Panda

Adult

IUCN: EN
CITES: I
CPS: II
CRL: VU

BL: 51–73 cm
TL: 31–48 cm
WT: 2.5–6 kg

Distribution

The distribution of red pandas includes forested regions of Nepal, northern India, Bhutan, Myanmar, and southwest China. In China, the red panda is found in the Minshan, Qionglai, Liangshan, Daxiangling, and Xiaoxiangling Mountains of western Sichuan, as well as northwest Yunnan Province and southeast Tibet AR.

Appearance

The red panda is a small raccoon-like carnivore with a reddish-brown-colored pelage and a long tail that is more than half of the body length. The color of the legs and lower part of the body are darker than the upper body. The tail is the most distinctive part of this species because it is thick and long and has distinct dark rings. The head is round with a short muzzle. The ears are erect, pointed, and large, with white hairs on the margin of the ear. The white hairs also cover the face part of cheeks and around the eyes and nose, making a dark facial mask. The western population along the Himalayas has a much paler pelage, especially around the face and head.

Habitat and Ecology

The red panda inhabits temperate forest habitat with dense bamboo understories and sometimes occurs at higher elevations near the tree line up to 3,600 m. Studies in Sichuan revealed that red

Track from a captive animal

pandas prefer microhabitats containing more fallen trees, which may help them to access new bamboo leaves without crossing the ground. The red panda diet is composed almost entirely of bamboo, such as young leaves and shoots (more than 95% of their total diet), fruit, roots, grasses, acorns, and lichens. Red pandas will also consume animals they can capture, including bird eggs, insects, and grubs. Studies also indicate available water is an important factor for red panda habitat selection. Red pandas are good tree climbers and are frequently found resting on tree branches. Little is known about their life history, but captive records show that the litter size is from one to four kits, and the young are sexually mature at 10 months of age.

Scat

Red panda scats are typically separate oval-shaped fecal pellets (3.5–4.0 × 1.7–2.5 cm). The pellets may be tied to each other with long plant fibers. Fresh scats are light to dark greenish, and old, dry scats are usually yellow to light brown. The pellets normally contain undigest-

Fresh scats

ed leaves, debris of small branches of bamboo, and other plant fibers.

Tracks

Red pandas rarely leave clear tracks, as the ground is well covered with moss and leaf litter, but their track can occasionally be found after a snowfall during winter. However, their footprints are normally blurry and difficult to identify on the snow because of their hairy footpads. The fore print (4-5 × 5-6 cm) is smaller than the rear print (4-5 × 8-9 cm).

BADGERS, MARTENS, OTTERS, AND WEASELS // HUAN, DIAO, SHUI TA AND YOU

General Description

Weasels and mongooses are small-bodied carnivores that are widely distributed across southwest China and often in abundant numbers. Many weasel and mongoose species are diurnal and frequently encountered in the wild. They are agile and effective predators on small vertebrates. Their slender body enables them to hunt in burrows, a unique niche among the carnivore families. Here we present their typical track and sign as a group since they share common traits and are usually indistinguishable from each other.

Tracks

The tracks of weasels and mongooses are rarely found in the wild, primarily because of their light body weight and small imprint. Tracks of weasels in temperate habitats are occasionally observed after snowfalls, but normally without clear structure or shape. Small weasels and mongooses frequently exhibit a jumping rather than a walking pace during fast movements.

Scat

Weasel and mongoose scats are typically found on an elevated surface, such as rocks and logs, and are deposited along their trails. Often, there are multiple scats of varied age at the same location. Typical scats of an animal diet are normally a single twisted, long, thin rope with a pointed tail at one end. Fresh scat is normally brown to black with a moist smooth surface, whereas older scats may turn gray to white. Scats often contain rodent hair, bone, and bird feathers.

Old scat of Siberian weasel

Fresh scat of Siberian weasel

Asian Badger

Meles leucurus // Ya Zhou Gou Huan

Family group

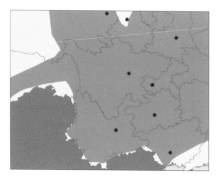

IUCN: LC	BL: 49.5–70 cm
CITES: NL	TL: 13–20.5 cm
CPS: NL	WT: 3.5–9 kg
CRL: NT	

Distribution

The distribution of the Asian badger includes large areas of East Asia, extending to central Asia and eastern Russia. The Asian badger was formerly considered a subspecies of the Eurasian badger, *M. meles,* which has been divided into three species with minimum overlap (the European badger, *M. meles*; the Asian badger, *M. leucurus*; and the Japanese badger, *M. anakuma*). In our focus area, we find the species to be more patchily distributed than indicated by IUCN. We know of populations distributed in the higher elevations of Gansu, Sichuan, and Shaanxi Provinces, as well as eastern Qinghai Province and eastern Tibet AR. Although being sympatric with the northern hog badger, *Arctonyx albogularis,* Asian badgers typically occupy higher elevations and more arid habitats.

Appearance

The Asian badger has a shape and color pattern similar to those of the northern hog badger, with a long conical head and piglike snout. The difference between the Asian badger and the hog badger is an external black nose pad, with hair evident between the nose pad and the upper lip. Another different characteristic is that the throat of the Asian badger is black, compared with the white throat of the hog badger. The Asian badger also has a distinctive facial pattern with narrower dark strips across its eyes and ears, resulting in the face appearing whiter than that of the northern hog badger. Old individuals may have a nearly white pelage across the body.

Habitat and Ecology

Very little is known about the ecology of Asian badgers. They inhabit both forested and open habitats, such as the alpine and subalpine meadow and scrublands, up to 4,500 m elevation in western China. The Asian badger is an omnivore, with a broad diet that can include invertebrates such as insects and earthworms, reptiles and amphibians, plant roots, mushrooms, and small mammals such as pikas and rodents. Asian badgers normally live in groups, sharing burrow systems called sets. Mating occurs in December and early January, and females produce one litter per year.

Scat

Typical scats of Asian badgers are shaped like a curved or curled rope with a diameter of 1.0–2.0 cm. The scats normally contain a large amount of earth consumed while digging for food. Old scats are typically dry and yellowish to black in color. The scats may also vary in shape and form according to diet. Scats composed of plant items are smooth without twisting. Scats composed of meat include fur and bones and are normally twisted, with very pointed ends. Undigested beetle shells are frequently found in their scats. The badgers usually have latrine sites near their burrows and piles of accumulated scats can be found.

Tracks

The Asian badger track has five toes with claws, although the fifth is frequently ab-

sent. The imprint has a single trapezoid plantar pad, with two lobes on the front edge and hollow rear edge. The tracks of the front feet are wider, with a longer claw print, than those of the hind feet.

Other Sign

Like other badger species, Asian badgers build a complex burrow system with multiple chambers underground. The burrow entrance can be as wide as up to 30–45 cm. Latrine sites, where the badgers defecate regularly, can be found near the burrow entrance.

Scats at latrine site

Dry scat

Burrow entrance

Track

Chinese Ferret-Badger

Melogale moschata // You Huan

Small-Toothed Ferret-Badger

Adult

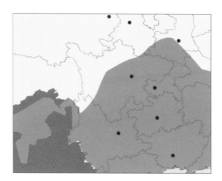

IUCN: LC
CITES: NL
CPS: NL
CRL: NT

BL: 30.5–43 cm
TL: 11.5–21.5 cm
WT: 0.5–1.6 kg

Distribution

The Chinese ferret-badger is distributed in the lowlands of central and Southeast Asia. In the region of this guide its distribution includes lower elevations of all provinces and ARs with the exception of Qinghai and Gansu Provinces and Tibet AR.

Appearance

When compared with the sympatric hog badger and Asian badger, the ferret-badger is much smaller and slenderer, with a long brushy tail, and the overall body shape is similar to that of a large weasel or marten. The pelage of the Chinese ferret-badger is similar to that of the other badgers in southwest China, with a black-and-white color pattern. They have whitish cheeks, a heart-shaped white patch between the eyes, and a triangle of black around each eye that looks like spectacles. The nose and the forehead are dark, and a white strip on the top of the head extends to the nape. The tail is proportionally much longer than that of other badgers, about half of the head and body length.

Habitat and Ecology

The Chinese ferret-badger is a subtropical mammal found in forest, shrubland, grassland, and agricultural land near human settlements. Chinese ferret-badgers are nocturnal and omnivorous. Their chief food is soil invertebrates such as earthworm and insects, as well as fruits and seeds. They also feed on small mammals, reptiles, and bird eggs. Little is known about their natural history; reports from different areas indicate the mating season is around March, and females give birth between the months of May and December to a litter of one to four kits.

Scat

The scat of Chinese ferret-badgers is similar to other small-omnivore scat (diameter <1 cm). Fresh scats are moist, with a color of black to dark gray. Like those of other badgers, the scats of Chinese ferret-badgers usually contain a large proportion of earth and remains of undigested insect shells, fruit seeds, and other plant materials.

Tracks

The track of a Chinese ferret-badger is normally smaller than 5 cm (including the claw marks), with five toes and claws; the first (inner) toe is the smallest and nearest to the plantar pads. The imprints of the plantar and metacarpal pads are fused and nearly round, with a hollow on the outer side.

Northern Hog Badger

Arctonyx albogularis // Zhu Huan

Hog Badger

Adult

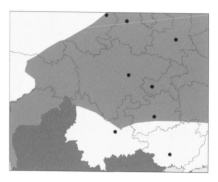

IUCN: LC
CITES: NL
CPS: NL
CRL: NT as *A. collaris*

BL: 32–74 cm
TL: 9–22 cm
WT: 9.7–12.5 kg

Distribution

The distribution of the northern hog badger is from northeast India through southern China. It was formerly recognized as subspecies of the greater hog badger, *A. collaris*, which is distributed across mainland Southeast Asia. In our focus area, the northern hog badger is distributed in Sichuan, Yunnan, and Guizhou Provinces; southern Qinghai, Gansu, and Shaanxi Provinces; eastern Tibet and Guangxi ARs; and Chongqing City.

Appearance

Northern hog badgers are medium-sized carnivores with a relatively long conical head and a piglike snout. Their most distinctive characteristic is the coat pattern, with a mostly white face, neck, throat, and ears, with two dark strips extending from the nose, over the forehead, and across the back of the neck. The hog badger's feet, legs, and belly are mostly dark brown or black, with a gray or light brown back. The tail is brushy and white.

Habitat and Ecology

The northern hog badger occupies forested hills and mountains and up to alpine meadows and rocky areas in southwest China from 200 to 4,400 m elevation. It is active both day and night and is commonly seen around human agriculture and residences. North-

Adult female and offspring

ern hog badgers are omnivores; their diet includes tubers, plant roots, earthworms, snails, insects, and, occasionally, rodents. They are good at digging with their strong legs and claws. Northern hog badgers are prey species of some larger predators like leopards and brown bears in southwest China. Females give birth in their burrow in February and March, and the litter size is from two to four. Animals near humans may cause substantial damage to crops.

Scat

The scats of northern hog badgers typically appear as a curved or curled rope with a diameter of 0.8–1.5 cm. The scat may sometimes break into two or three segments. The shape of the scat will vary with diet. Diets composed of vegetation,

Fresh scat

Digging pit

fruits, and seeds result in smooth surfaces, with little or no twisting. A meat diet results in scats containing fur and bone that are extremely twisted, with pointed ends. Normally, the scats contain large amounts of dirt consumed while digging for food. Beetle shells are also frequently found in their scats. Fresh scats are usually moist and dark brown or black in color, whereas old ones are dry and dark gray. Scats often accumulate at den sites and at the burrow entrance.

Tracks

Northern hog badgers leave typical badger tracks of five toes with claw imprints. The innermost toe is the smallest, and the track has a single wide metacarpal pad. The track for the front paw is much larger, with longer claws, than that of the hind track (around 6.5 × 4.5 cm for front feet and 5 × 4.5 cm for hind feet).

Other Sign

Northern hog badgers leave numerous digging sign as they forage. These digging sign are typically shallow pits with disturbed soil and leaf litter. Such pits are usually indistinct from the sign left by other mammals of similar body size (e.g., the Asian badger) or pheasants, which also dig for food.

Old scat

Beech Marten

Martes foina Shi Diao

Stone Marten

Adult

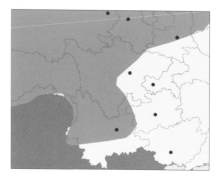

IUCN: LC
CITES: III
CPS: II
CRL: EN

BL: 34–48 cm
TL: 22–33 cm
WT: 0.8–1.6 kg

Distribution

The beech marten is widely distributed in Europe and Asia, from western and central Europe to mid-Asia and then extending into East Asia. Within the range of this guide, beech martens are found in western Yunnan and Sichuan Provinces, Gansu and Qinghai Provinces, northern Shaanxi Province, and the Tibet AR.

Appearance

Beech martens are medium-sized weasels with wooly fur that is dark brown to chocolate colored. The head is normally paler than the body, with a distinctive broad white patch on the throat that extends to the chest. There is usually a small dark central patch or spot within the throat patch. The lower parts of the four legs are darker than the body. The tail is brushy and about half the head and body length.

Habitat and Ecology

Beech martens are normally found upon the Tibetan plateau in open areas up to 4,600 m elevation. They are also reported to use forest edge and brushland at lower elevations. They are diet generalists: small mammals such as rodents and pikas are their main prey, but they will also eat fruits and berries when available, and the consumption of birds, reptiles, and insects has been reported. The breeding season is midsummer, and a normal litter is three or four kits. Beech martens have been widely hunted and trapped in western China for their fur.

Scat and Tracks

No specific information is available for this species.

Yellow-Throated Marten

Martes flavigula // Huang Hou Diao

Adult

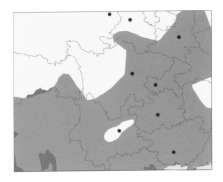

IUCN: LC
CITES: III
CPS: II
CRL: NT

BL: 52–72 cm
TL: 39–52 cm
WT: 1.3–3 kg

Distribution

Yellow-throated martens are widely distributed from eastern Russia through China to Southeast Asia and the northern Indian subcontinent. Within the range of this guide they are found in all provinces and ARs except for Qinghai Province.

Appearance

Yellow-throated martens are large weasels with a remarkably long tail, which is up to 70%–80% of its body length. This carnivore has a distinct color pattern, with the head, nape, hind limbs, and tail black to dark brown. In contrast, its throat, shoulder, chest, and upper forelimbs are bright yellow to golden. Its chin and cheeks are whitish. Yellow-throated martens are frequently active in pairs during the day and move quickly and agilely through the forest.

Habitat and Ecology

Yellow-throated martens are generalists that are found across a wide range of habitats, including multiple forest types, shrubland, and plantation forest from 200 to 3,500 m elevation. Martens are primarily diurnal, with a very broad diet, including small mammals, birds, eggs, frogs, reptiles, insects, and fruits. They are good climbers and frequently forage in trees. Yellow-throated martens have been reported to hunt for small ungulates (e.g., Reeves's muntjac, forest musk deer) in pairs or family groups. They sometimes consume honey in native and commercial bee hives,

Adult pair

for which they are called "honey dog" in southwest China.

Scat

Scats of yellow-throated martens vary according to the diet. Fresh scats with vertebrate animal remains (e.g., hairs, feathers, and tiny bones) are usually moist with smooth surfaces, have little twisting, and are segmented. Scats are found mostly on elevated surfaces along animal trails, such as logs, stumps, and rocks. They are distinct from scats of other sympatric small weasels by their larger size (diameter up to 1 cm). Bee remains can be found after a marten feeds on bee nests.

Tracks

Tracks of the yellow-throated marten are occasionally found on soft wet ground or snow. The track has five toes with claw marks, which is typical of large mustelids and unlike the felids and canids (i.e., cats and dogs), which leave four-toe tracks. Their tracks (2.8–3.4 × 2.1–2.8 cm) are distinct from those of other sympatric small weasels (e.g., Siberian weasel and yellow-bellied weasel) by the larger size and are distinct from otter tracks by the presence of claw marks. The size of yellow-throated marten tracks is similar to that of badgers (e.g., ferret badger and hog badger), but the badger tracks have distinctive long claw marks. The marten track usually shows a fast jumping or running pattern, in contrast to the typical slow walking pattern of badgers.

Fresh scat

Scat after feeding on a bee nest

Old scat

Asian Small-Clawed Otter

Aonyx cinerea / Xiao Zhua Shui Ta
Oriental Small-Clawed Otter, Small-Clawed Otter

Adult

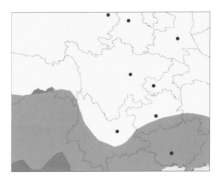

IUCN: VU
CITES: II
CPS: II
CRL: EN

BL: 40–61 cm
TL: 29–35 cm
WT: 2–4 kg

Distribution

The Asian small-clawed otter has a broad distribution that includes South Asia and Southeast Asia and extends into southern China. Within our focal area, the species was widely distributed across the southern provinces, but currently it is reported at only a handful sites in western Yunnan Province.

Appearance

The Asian small-clawed otter is the smallest otter species in the world. It is uniformly dark brown on the head, back, legs, and tail and grayish white on the lower face, throat, and upper chest. The base of tail is very thick but tapers quickly. Its claws are vestigial, and its feet are partially webbed. In the wild, it may be indistinguishable from Eurasian otters.

Habitat and Ecology

Small-clawed otters are found in freshwater wetlands habitat, such as small streams, ponds, rice paddies, marshes, swamps, and mangroves. They coexist with other otter species but prefer small water bodies up to 2,000 m elevation. Their diet includes freshwater crabs, snails, fish, frogs, and insects. They forage in groups of 12–15 individuals and leave scent marks frequently as territorial markers. Females may have two litters a year, and the average litter size is four kits.

Scat

Otter scats may range from a loose, liquid patty to tubular scat (diameter <1.0 cm), depending on diet. They are similar in shape and size to that of the Eurasian otter.

Tracks

Five toes normally show; the palm pads are fused together. The web between the toes may show up on sand or other soft surfaces. The claws are not visible. Tracks are similar in shape and size to those of the Eurasian otter.

Eurasian Otter

Lutra lutra // Shui Ta

Common Otter, Old World Otter

Adult

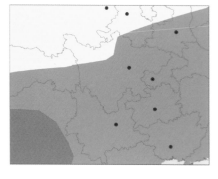

IUCN: NT
CITES: I
CPS: II
CRL: EN

BL: 49–84 cm
TL: 24.3–44 cm
WT: 2.5–9 kg

Distribution

The Eurasian otter is one of the most widely distributed mammal species in the world. It occupies most of the Eurasian continent, parts of North Africa, and some of the islands in Southeast Asia. Until recently, all the provinces and regions in our focus area were within the range of Eurasian otters; however, their current distribution in southwest China might be highly fragmented. Confirmed records during the past 10 years are reported from only a handful sites in southern Qinghai and Shaanxi Provinces, western and northern Sichuan Province, northwest Yunnan Province, and eastern Tibet AR.

Appearance

Eurasian otters have a thick and dense fur that is brown to dark brown on the body, legs, and tail. The fur of the ventral area and throat is paler than the back. Their tail is conical, thick, and muscular. Their ears are small. The legs look relatively short compared with the body, with webbed feet and well-developed claws.

Habitat and Ecology

Eurasian otters live in freshwater habitat, such as rivers, lakes, marshes, swamps, and rice fields, and they prefer shallow waters. Their principle diet is fish, but they may occasionally consume frogs, birds, crustaceans, lagomorphs, and rodents. Eurasian otters are solitary, and their activity is either nocturnal or crepuscular. They are territorial and mark their territories with scent from anal glands and scats. They reach maturity at 2 or 3 years; gestation lasts 63 days, and females give birth to two to three kits. Eurasian otters in China have experienced a dramatic decline and range constriction during the past half century due to extensive harvest for their fur and body parts (e.g., livers for use in traditional Chinese medicine).

Scat

Otter scat may range from a loose, liquid patty to tubular scat, depending on the diet type (5–9 cm in length and diameter <1 cm). Scats are frequently found on top of rocks near water, and the otters may have latrines that they regularly visit. Fresh scats have a strong, stinking odor and usually contain numerous fragments of undigested fish bones and scales and remains of other prey like crab shells.

Tracks

Tracks show five toes with claws, with relatively round shaped palm pads. The web between toes may be visible on wet sand or other soft surfaces. Occasionally, the mark of the tail being dragged is visible between the foot tracks.

Scats

Fresh tracks

Latrine

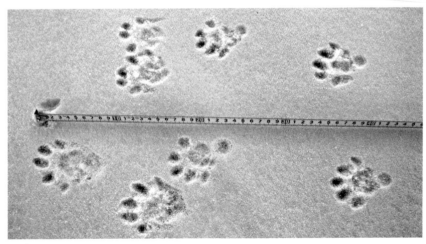

Track

Mountain Weasel

Mustela altaica // Xiang You

Altai Weasel, Pale Weasel

Adult with summer coat

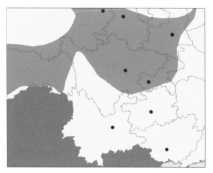

IUCN: NT	BL: 10.5–27 cm
CITES: III	TL: 6.6–16.2 cm
CPS: NL	WT: 80–340 g
CRL: NT	

Distribution

The mountain weasel is distributed in eastern Russia and northwest China and into central Asia. In the area of this guide, the species is found in the alpine areas of western and northern Sichuan, eastern Qinghai, and southern Gansu Provinces, as well as the eastern Tibet AR.

Appearance

Mountain weasel coloration is similar to that of the Siberian weasel, but its body size is smaller, the dark facial mask is normally absent, and the tail lacks a black tip. Another distinguishing characteristic is that the ventral fur is yellowish or white and sharply contrasts with the dorsal fur. There is a visible dividing line between the dark dorsal side and the paler ventral side. The feet are also white and contrast with the darker back and legs. Tail length is about one-half to two-thirds of the head and body length. The summer coat is shorter, coarser, and darker than the winter coat. The dividing line between the dorsal and ventral sides is more distinct in the summer coat. The tail is much thicker and brushy during winter. Males (230–340 g) are typically larger than females (80–230 g).

Rocky habitat and weasel with winter coat

Habitat and Ecology

Mountain weasels inhabit alpine meadows and often appear in rocky slopes between 2,500 and 4,500 m elevation in southwest China. They prey primarily on pikas and voles, as well as birds, lizards, and insects. Mountain weasels are also reported to eat berries and attack domestic fowl. Mountain weasels can climb and swim well. When alert, they may stand on their hind feet with their whole body upright. They are reported to be polygynous breeders, with a gestation of 35–50 days. The young are observed in the summer months.

Siberian Weasel
Mustela sibirica   Huang You
Yellow Weasel

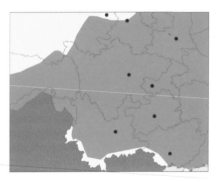

Adult with winter coat

IUCN: LC
CITES: III
CPS: NL
CRL: LC

BL: 22–42 cm
TL: 12–25 cm
WT: 0.5–1.2 kg

Distribution

The Siberian weasel is widely distributed in eastern Russia, Japan, Korea, and a large part of China, and its range extends into Southeast Asia. The Siberian weasel is distributed across all the provinces and regions included in this guide except for the western plateau regions of Qinghai Province and Tibet AR.

Appearance

The Siberian weasel is generally orange brown, with a black or dark brown mask on the face and white on the muzzle and chin. The ventral side is slightly paler than the dorsal side. Its summer coat is darker, and the winter coat is thicker and paler. The tail is brushy, usually with a dark tip, and about half the head and body length.

Habitat and Ecology

Siberian weasels are found in all habitats across a broad elevation range from the coast to 5,000 m, including primary and secondary forest, dry steppe forest, river valley, and cultivated areas around villages. They are also frequently found in the urban areas. Their generalist diet includes rodents, shrews, birds, amphibians, invertebrates, berries, nuts, and sometimes domestic fowl. They are capable of chasing and hunting prey (e.g., rats) in their burrows. Siberian weasels are nocturnal and crepuscular and normally solitary. The breeding season is March and April, and the females give birth in May to a litter of five or six kits.

Steppe Polecat
Mustela eversmannii // Ai You
Polecat, Steppe Weasel

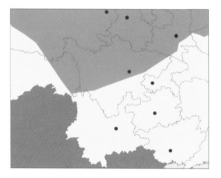

Adult pair

IUCN: LC	BL: 31.5–46 cm
CITES: NL	TL: 9–20 cm
CPS: NL	WT: 0.5–1.2 kg
CRL: VU	

Distribution

The steppe polecat is widely distributed from central Europe to East Asia. Within the range of this guide, it is found in the northern provinces of Qinghai, Gansu, and Shaanxi, as well as western Sichuan Province and eastern Tibet AR.

Appearance

The steppe polecat is a large weasel, with a distinctive color pattern. The throat, feet, tail, and ventral surface are black. The dorsal side is yellowish with black-tipped guard hairs, which may appear as a broad black dorsal strip on the center of the back. The face is whitish to pale yellow with a dark mask. The upper edge of the ear is white. The tail is brushy, with a length less than half of the head and body length. The winter coat is much paler, sometimes nearly white on the head and flanks, and thicker than the summer coat.

Habitat and Ecology

Steppe polecats use open steppe habitat rather than forest. In the Qinghai-

Tibet Plateau they feed mainly on pikas, so their population numbers fluctuate with the density of their prey. They also feed on rodents, birds, reptiles, and insects, when available. They normally hunt the small-mammal prey in their burrows. Steppe polecats are mainly nocturnal but can be active during the day. The breeding season is early spring, and females give birth to a litter of 4 to 10 kits. Adults are typically solitary, although they are occasionally observed in pairs or mother-offspring groups.

Exiting a burrow with its prey

Front view of adult

Yellow-Bellied Weasel

Mustela kathiah // Huang Fu You

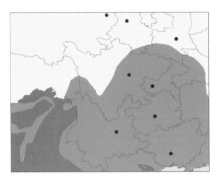

Adult with summer coat

IUCN: LC
CITES: III
CPS: NL
CRL: NT

BL: 20.5–33.4 cm
TL: 6.5–18.2 cm
WT: 168–250 g

Distribution

The yellow-bellied weasel is distributed in central and southern China, and its range extends into Southeast Asia. Within our focus area the species is reported in all the eastern provinces and ARs at lower elevations.

Appearance

The yellow-bellied weasel is a small weasel, with relatively dark brown fur on its back and tail, orange or yellowish fur on its underparts, and white on the chin. There is a conspicuous straight dividing line on the side of its neck and body between the dark dorsal side and the pale ventral side. Yellow-bellied weasels have white patches on their feet. The tail length is about half of the head and body length, proportionally short compared with that of the Siberian weasel and mountain weasel. The winter coat is denser, longer, and paler than the summer coat.

Habitat and Ecology

Very little is known about the ecology of yellow-bellied weasels. They are typically found below 2,000 m elevation and in various forest habitats. They may also inhabit degraded forests, plantations, and shrublands near humans. Yellow-bellied weasels feed on various small prey, including rodents, birds, lizards, frogs, and insects. They are nocturnal and solitary and appear to be territorial. They build dens in the ground, and the breeding season is typically during spring.

CIVETS, LINSANG, AND MONGOOSES
LING MAO, LIN LI, AND MENG

General Description

Civets and linsangs were both placed in the family Viverridae, but the Asian linsangs have recently been placed into a separate family, Prionodontidae. Despite the taxonomic changes, both civets and linsangs are elusive animals and are rarely observed in the wild. When compared with other small carnivore groups (i.e., cats, foxes, weasels, and badgers), civets and linsangs leave many fewer signs of their presence. Their omnivorous diet results in fast degradation of their scats, especially within their typical moist and warm environments, and the defecation behavior of some civet species also makes their scats difficult to locate. Multiple species within this group may co-occur in this region, such as the masked palm civet (*Paguma larvata*) and large Indian civet (*Viverra zibetha*) in the subtropical forests in western Yunnan and southern Sichuan Provinces. The tracks and scats of sympatric civets may not be distinguishable to species.

Here we present the typical tracks and sign of the civet-linsang group using the masked palm civet, the most widely distributed civet species within the region, as a representative.

Track of masked palm civet

Fruit remains on ground after being eaten by masked palm civet

Tracks

The tracks of civets and linsangs are rarely found in the wild, primarily because of the heavy leaf litter and wet environment. Fresh tracks of civets are occasionally observed on soft or muddy ground. Five toes are registered on the track, and the toes appear close to the plantar pads. The prints of the plantar pads are round for the forefeet and a long triangle for the rear feet.

Scat

Civets and linsangs are omnivorous, and their diet includes small mammals, other vertebrates, and plant material. Wild fruits and berries form a significant proportion of the diet of multiple civet species, especially during the fruiting seasons. Accordingly, the scats of civets and linsangs vary broadly in shape, color, and components depending on their diet. Typical civet scats after a meal of fruits and berries are in segments with a large number of undigested seeds. The scats are normally a twisted long rope with a pointed tail at one end, similar to that of larger weasels and martens. Some civet species, such as the masked palm civet, frequently defecate near water (usually small running streams) or even in shallow water. These scats are easily flushed by the streams and result in low detection rates for their scats.

Other Sign

Most civets are nocturnal, and they usually rest in well-sheltered dens either alone or in family groups during the day. These dens are in natural small caves, rock crevices, or sometimes old dens abandoned by other animals, such as porcupines. Civet hairs are normally found on the rocks or tree roots at the den entrance.

Civets and linsangs are agile tree climbers and will readily feed on wild fruits and berries in trees or shrubs. A common sign is fallen fruit remains under these trees, but such feeding sign may be indistinguishable from the sign left by other foraging mammals (e.g., yellow-throated martens, macaques) and large birds.

Common Palm Civet

Paradoxurus hermaphroditus // Ye Zi Li

Mentawai Palm Civet, Toddy Cat

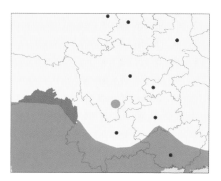

Front view of adult

IUCN: LC
CITES: III
CPS: NL
CRL: VU

BL: 47–57 cm
TL: 47–56 cm
WT: 2.4–4 kg

Distribution

The common palm civet is widely distributed in South and Southeast Asia, extending into southern China. Within the range of this guide, it is found in southern Yunnan Province and southern Guangxi and Tibet ARs. There were scattered reports from southern Sichuan Province, but with no recent confirmed records.

Appearance

The size of the common palm civet is similar to that of the masked palm civet, but it has a thinner and proportionally longer tail and a spotted coat. The fur is yellowish gray or light brown. It has at least five longitudinal rows of dark spots on the back, which merge into dorsal stripes near the rump. The fur on the tail, feet, and face is black, and the tail is equal in length to the head and body.

Habitat and Ecology

Common palm civets live in tropical and subtropical montane forests, as well as degraded and anthropogenic (e.g., plantations and croplands) habitats. They are opportunist omnivores and feed heavily on fruits, nuts, and berries. They also consume animal food such as arthropods and small vertebrates, and occasionally, they kill domestic poultry. Common palm civets are solitary and nocturnal to crepuscular and appear to be territorial, with notable arboreal activities. They are thought to be aseasonal breeders, with a gestation of about 60 days, and females give birth to an average litter size of three kits. As with the masked palm civets, they are also widely poached in southern China for meat.

Large Indian Civet

Viverra zibetha // Da Ling Mao

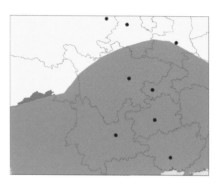

Front view of adult at night

IUCN: LC
CITES: III
CPS: II
CRL: VU

BL: 50–95 cm
TL: 38–59 cm
WT: 3.4–9.2 kg

Distribution

The global distribution of this animal includes most of Southeast Asia and the northeast region of South Asia. In the focus area of this guide, the species was historically reported in most of the southern provinces, but its current range is poorly known. Recent confirmed records are from only a handful fragmented areas in southern Yunnan and Sichuan Provinces and southeast Tibet AR.

Appearance

This is the largest civet species in Asia, with a robust body, thick neck, and banded, thick tail. The fur is gray to grayish brown; the spots on the body are indistinct and interconnected, which makes the spots look blurry. The most distinctive characteristic of large Indian civets is the clear black and white stripes on the neck and the dark rings on the

tail, with brown between the black rings. The tail is longer than half of the head and body length.

Habitat and Ecology

In China, large Indian civets are found only in dense tropical and subtropical forests, and little is known about their natural history and ecology. Information from studies outside of China indicate they are mainly carnivorous and consume birds, frogs, snakes, small mammals, eggs, and

Back and tail view of adult

fish and only occasionally consume fruits and roots. Large Indian civets are territorial and are more active during the day. As aseasonal breeders, they can have two litters a year, with each litter composed of one to five young. They are severely poached in southern China for food, and many regional populations may have been eliminated.

Masked Palm Civet
Paguma larvata // Hua Mian Li
Himalayan Palm Civet, Gem-Faced Civet

Front view of adult

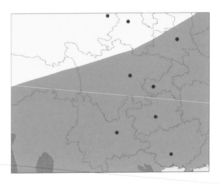

IUCN: LC
CITES: III
CPS: NL
CRL: NT

BL: 40–69 cm
TL: 35–60 cm
WT: 3–7 kg

Distribution

The masked palm civet has the largest geographic range among all civet species. Its distribution ranges from central and southern China (partly extending northward into northern China near Beijing along the Taihang Mountains), westward along the Himalayas, and southward into Southeast Asia. In the focus area of this guide, the species occurs in all provinces and ARs with the exception of Qinghai Province and most of Gansu Province.

Appearance

Masked palm civets are large civets with a robust body, short limbs, and long tail. The base color varies within the region; they are normally rusty brown to brownish gray, sometimes pale golden, with darker fur on their legs and tail tip. The ventral side is normally paler than the dorsal side and flanks. Masked palm civets do not have spots or stripes on their body and tail, which is distinctive from other sympatric civets such as the

Den

common palm civet, large Indian civet, and small Indian civet. The most distinguishing characteristic of the species is the black-and-white mask on the face, with the black rostrum, a white stripe on the middle of the head extending to the nape, small white spots under the eyes, and larger white patches above the eyes. The tail of the masked palm civet is thick and long, with the length longer than half of the body length.

Habitat and Ecology

Masked palm civets can inhabit all forest types, from primary evergreen forest to secondary deciduous and coniferous forests, and they have frequently been found in agriculture areas. They may occur across a broad elevational range from sea level to above 3,500 m in the high mountains of southwest China. Masked palm civets are omnivores, and their diet contains fruits, berries, plant roots, birds, rodents, and insects. They have also been reported to prey on domestic chickens

and waterfowl and frequently scavenge on carcasses. Masked palm civets are good climbers and forage for fruits such as wild cherries and Chinese bayberries, for which they are called "fruit civets" in southern China. They are nocturnal animals and rest in burrows during the day. Masked palm civets are usually solitary, but groups of 2–10 individuals are ob-

Side view of adult

Fresh scats

Scat

served. They reach sexual maturity in 1 year, and the gestation period is 70–90 days. In the mountains of southwest China, masked palm civets may reduce their activities during the winter, as extensive camera trapping in this region shows a near absence of detections from early December through March. Masked palm civets may damage fruit plantations, for which they are commonly killed during the fruit harvest seasons. They are also widely poached in southern China for meat, although they are an important intermediary host of various zoonotic viruses (e.g., the SARS virus). Captive farming of this species is also common in China, and small populations within the region may contain escaped individuals.

Scats with plant and animal remains

Small Indian Civet

Viverricula indica // Xiao Ling Mao

Oriental Civet

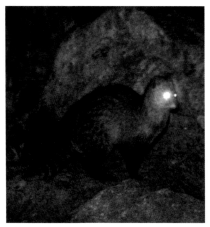

Adult

IUCN: LC	BL: 50–61 cm
CITES: III	TL: 28–39 cm
CPS: II	WT: 1.6–4 kg
CRL: VU	

Distribution

The global distribution of the small Indian civet includes most of Southeast Asia and South Asia. In the focus area of this guide, the species is distributed at low elevations in most of the provinces and regions except for the high plateau of Qinghai Province and most of the Tibet AR.

Appearance

The small Indian civet is a medium-sized civet with a slender body, a spotted coat, and a thick tail with conspicuous dark rings. Its coat is gray to grayish brown, with darker colored feet. Black or dark brown spots run in longitudinal rows along the body, and the spots on the back may form five to seven parallel dorsal stripes spreading from shoulder to rump. There are black and white rings on the tail, and the tail is always white on the tip.

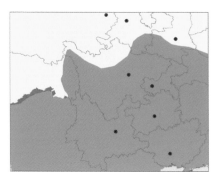

Habitat and Ecology

Small Indian civets can be found in grassland, scrubland, secondary forest, and agricultural lands. They are omnivores, and their diet includes small mammals, insects, earthworms, birds, reptiles, eggs, crustaceans, and snails, as well as fruits and buds. They occasionally kill domestic poultry. They are solitary, and their activity is primarily nocturnal to crepuscular. Very little is known about their social organization or breeding.

Scat and Tracks

No specific information is available for this species.

Spotted Linsang
Prionodon pardicolor // Ban Lin Li

Adult

IUCN: LC
CITES: I
CPS: II
CRL: VU

BL: 35-45 cm
TL: 30-40 cm
WT: 0.55-1.2 kg

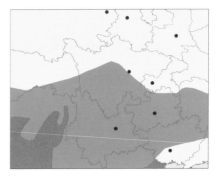

Distribution

The spotted linsang is distributed throughout Southeast Asia, and its range extends into southern and southwest China. For the region covered by this guide, the species is distributed in Yunnan, Guizhou, and southern Sichuan Provinces, as well as Guangxi and southern Tibet ARs.

Appearance

The spotted linsang was previously considered a small civet under the family Viverridae, but recent studies have placed the two Asian linsang species (spotted linsang, *P. pardicolor,* and banded linsang, *P. linsang*) in their own family, Prionodontidae. The spotted linsang has a slimmer body and longer neck than similar civet species within the range. Its fur is light yellowish gray with dark solid spots. The round dark spots roughly array in parallel rows along the body. The tail is as long as the head and body length, with black rings and a white tip.

Habitat and Ecology

Spotted linsangs primarily inhabit tropical and subtropical evergreen broadleaf forest. They can also be found in grassland, scrub, or disturbed forest habitats. Their principle diet includes small vertebrates such as rodents, shrews, frogs, and reptiles; bird eggs; and some berries. They are arboreal, solitary, and nocturnal predators. Spotted linsangs breed from February to August, and females produce litters with two to four young.

Scat

No specific information is available for this species.

Tracks

Linsang tracks are rarely observed in the field because of linsangs' arboreal habits. The track is small (2.5 × 3 cm) and may be occasionally found on the ground in soft, wet soil. The track of the forefoot shows five toes with claws, with the first toe much lower and nearer to the palm pad and the third and fourth toes more forward. The hind foot normally shows a print of four toes with claws.

Crab-Eating Mongoose
Herpestes urva // Shi Xie Meng

Adults

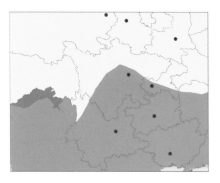

IUCN: LC	BL: 36–52 cm
CITES: III	TL: 24–33.6 cm
CPS: NL	WT: 1–2.3 kg
CRL: NT	

Distribution

The crab-eating mongoose is distributed from Southeast Asia through China's southern regions. Within the range of this guide, the species is found in southern Sichuan and Chongqing City, as well as Guizhou and Yunnan Provinces and Guangxi AR.

Appearance

The crab-eating mongoose is a large mongoose species with stocky body and long tail. Its fur is grizzly pale gray to dark grayish brown, with a pale brushy tail and dark legs. The mouth and chin are white, and a white stripe extends to the neck. This white stripe with extended long hairs is the most distinctive characteristic of the species.

Adult

Habitat and Ecology

Very little is known about the ecology of crab-eating mongooses. They are normally found in evergreen forest near streams and rice fields from low elevations up to 2,000 m. They normally forage along streams for fish, frogs, crabs, insects, and earthworms. They are crepuscular or diurnal animals and appear to be solitary predators but are also frequently found in pairs.

Scat

No specific information is available for this species.

Tracks

The tracks of crab-eating mongooses are normally found near water on mud or sand (4.0–4.5 × 3.5–4.5 cm). The prints of the forefeet normally present four toes with claws and palm pad; the hind feet may present five toes, and the first toe is in a much lower position but frequently indistinct and connected to the pad.

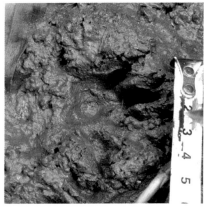

Forefoot track

Hind foot track

Javan Mongoose

Herpestes javanicus // Zhao Wa Meng

Adult with winter coat

IUCN: LC
CITES: III
CPS: NL
CRL: VU

BL: 25–37 cm
TL: 24–27 cm
WT: 0.6–1.2 kg

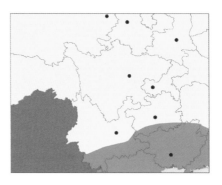

Distribution

The Javan mongoose and the small Indian mongoose (*H. auropunctatus*) were considered one species, but recent studies indicate they are two distinct species. The Salween River (Nujiang River in China) is presumed to be the geographic dividing line between the two species. The Javan mongoose is distributed across Southeast Asia and extends into southern China. Within our focus area the species is found in Guangxi AR and southern Yunnan Province.

Appearance

The Javan mongoose is the smallest mongoose species in Asia. The overall body shape is long and slender with a long, brushy tail, appearing similar to a large weasel. Its dorsal color is uniform brownish gray. Ventral fur is chestnut color. The fur on the cheeks and chin is ferruginous. The tail is equal in length to the head and body. The distinctive characteristic compared to the crab-eating mongoose is the smaller body size and the shape and look of tail; the tail of the Javan mongoose is less densely furred.

Habitat and Ecology

The Javan mongoose is found in dry forest, grassland, secondary forest, and agricultural lands. Its basic ecology is poorly known, but its generalist diet includes mainly insects, as well as rodents, birds, reptiles, frogs, fish, and fruits. Javan mongooses are diurnal and are aseasonal breeders, with a gestation of 50 days and one litter per year with two to six young.

Scat and Tracks

No specific information is available for this species.

BIG CATS // DA MAO

Clouded Leopard
Neofelis nebulosa // Yun Bao

Adult

IUCN: VU
CITES: I
CPS: II
CRL: CR

BL: 70–108 cm
TL: 55–91.5 cm
WT: 16–32 kg

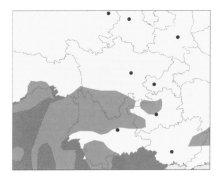

Distribution

The clouded leopard is distributed from the Himalayan foothills in Nepal through mainland Southeast Asia into southwest and southern China. This species historically had a wide distribution in China south of the Yangtze River. During the past 50 years its range in China has been severely reduced, and recent confirmed records are from only a handful sites in southern and western Yunnan Province and southeast Tibet AR; the species may still exist in southern Sichuan Province. The range shown is reduced from that indicated by IUCN on the basis of our local knowledge.

Appearance

The clouded leopard is a medium-sized cat with a long tail and relatively small head. The males are normally slightly larger than the females. Clouded leopards have longer hind legs than forelegs, resulting in an obvious arched hip and lower shoulder. Their pelage is light yellow to light gray, with large, irregular, oval-shaped black patches on the body. The ventral side is white with black spots. They have two broken black stripes along the center of the back down to the tail base and six longitudinal black stripes on the neck. Their legs have large black solid spots. The ears are round and black on the back. The tail is much longer than half of the head and body length, with black incomplete rings. A melanistic form (i.e., black) is occasionally reported, on which the base color of the body pelage is replaced by black or dark gray.

Habitat and Ecology

Clouded leopards are associated with tropical and subtropical forests and occupy both primary and secondary forests. They feed on civets, boars, macaques, small mammals, pheasants, and other birds. They are solitary and nocturnal. Clouded leopards are good tree climbers and swimmers, although they forage mainly on the ground. Females give birth to about three kittens.

Scat

Clouded leopard scats are normally ropelike with a diameter of 1.0–2.0

cm. The scats are typically segmented, and the last segment usually has one needle-shaped end. When they are sympatric with other small- to medium-sized cats, such as Asiatic golden cats, leopard cats, or marbled cats, the scats may be difficult to distinguish.

Tracks

The clouded leopard has a medium-sized cat track (4.5–5.5 cm across) with four toes registered on the track, and the track is more symmetrical than that of other large cats. The plantar pads in prints are nearly trapezoid shaped, with two lobes on the front edge and a hollow on the rear edge.

Scat

Leopard

Panthera pardus // Bao

Adult

IUCN: VU	BL: 91–191 cm
CITES: I	TL: 51–100 cm
CPS: I	WT: Male, 20–90 kg;
CRL: EN	Female, 17–42 kg

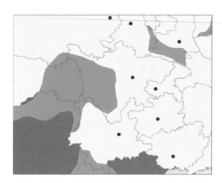

Distribution

The leopard is the most widely distributed felid species in the world, ranging across Eurasia and Africa. Within China the leopard range has been significantly reduced during the past half century to several fragmented areas. In our focus area the current records are from only southern Shaanxi, Gansu, and Qinghai Provinces; western Sichuan Province; southwest Yunnan Province; and southeast Tibet AR. The range shown is reduced from that indicated by IUCN on the basis of our local knowledge.

Appearance

Leopards are the largest spotted cats in China. Adult males are larger than females. They have a light brown to yellow or orange pelage, with conspicuous large black rosette-shaped spots on the back, tail, and flanks. The ventral pelage and the inner side of the legs are white. The spots on their head, legs, and venter are solid black dots or spots. The melanistic form (i.e., so-called black leopard or black panther) is also occasionally reported, especially in tropical or subtropical forests, on which the typical pelage background color of yellow is replaced with black or dark gray. Their ears are round and wide set on the head. The legs appear proportionally short. The thick tail is longer than half of the head and body length, with black rosettes. The rosette form (also called the ocelot form) of the Asiatic golden cat may appear similar to a smaller leopard, but leopards are much larger in size and have a relatively small head and longer and thicker tail, and the patterns of spots and stripes are different (see details in the description of the Asiatic golden cat).

Habitat and Ecology

Leopards are highly adaptable animals and occupy assorted habitats from tropical to temperate climates and along a large elevation range (up to 5,000 m). They are found in all habitats except desert or tundra. In western Sichuan and southern Qinghai Provinces, leopards have been observed overlapping with snow leopards within conifer forest, alpine shrub, and alpine meadow habitats above 4,000 m. Leopards are solitary animals, although mother-offspring groups of two to four individuals are also frequently reported. Adults are territorial, but the home ranges of nearby individuals may overlap with each other, and the males may hold a larger home range overlapping with multiple females. They are good tree climbers and may rest and hide the carcasses of their prey in big trees. They are primarily nocturnal but may hunt during the day as well. Leopards prey on many terrestrial vertebrates, including ungulates, large rodents, hares, primates, pheasants, and sometimes other small carnivores such as foxes and badgers. Although their preferred prey size is less than 50 kg, they are capable of killing larger prey such as wild boars, serows, white-lipped deer, and sambars. They may scavenge on animal carcasses as well. In western Sichuan and southern Qinghai, leopards may occur near human residences and sometimes prey on domestic animals, which causes numerous human-leopard conflicts. Leopards breed in February, and females have a gestation of 90–105 days, with a litter size of one to three kittens. The offspring will stay with the mother for 1–1.5 years before independence. The leopards in southwest China have suffered remarkable population decline and range reduction during the past century, primarily because of the high hunting and poaching pressure for their fur and their body parts (e.g., bones), which are used in traditional Chinese medicine. Retaliatory killing and poisoning caused by human-leopard conflicts also contribute to their decline.

Claw mark

Scat

Leopard scat is normally segmented with widely varying size (diameter of 1.5–4.0 cm). The last segment typically has a blunt start and a needle-shaped tail. The scat contains undigested animal hairs, skin, feathers, and sometimes plant fibers. Fresh scats have varied colors depending on the diet and have a strong stinking odor. Old, dry scats are typically loose and white. When leopards are sympatric with other large carnivores such as wolves, snow leopards, lynx, and free-ranging domestic dogs that feed on wild prey, their scats may be difficult to distinguish.

Tracks

Leopards leave one of the largest carnivore tracks (6.5 × 7.0 cm) within the region of this guide. The overall shape is nearly round, and each track registers four toes; normally, the second one is the largest. The toes of the forefeet are more spread out than those of the hind

track. The prints of the plantar pads are trapezoid shaped, with a short, round top margin and three lobes on the rear edge. The claw prints are absent.

Other Sign

Leopards leave claw marks on trees along trails within their territory. They usually stand on their hind legs, lift their body upward, and scratch on the tree trunk with their forefeet, resulting in clear marks. These claw marks are normally found 1.0–1.5 m aboveground on medium- to large-sized trees with coarse bark. The claw marks are more visible during winter and spring when the tree bark is dry and the forest understory is more open.

Like other cat species, leopards frequently create scratching pits with their hind legs, especially after they urinate or defecate. The scratching pit is normally a shallow depression with disturbed soil or leaf litter and is frequently accompanied by scats or shed hairs.

Fresh scat

Dry, older scat

Adult scratching

Scratching pit

Track

Snow Leopard

Panthera uncia // Xue Bao

Ounce

Adult male

IUCN: VU
CITES: I
CPS: I
CRL: EN

BL: 110–130 cm
TL: 80–100 cm
WT: 38–75 kg

Distribution

The snow leopard is distributed in the mountains from central Asia to the Qinghai-Tibetan Plateau. In our focus area, snow leopards are present in most of the high-elevation (>3,300 m) areas, including western Sichuan Province, the northwest part of Yunnan Province, southern Qinghai Province, western Gansu Province, and eastern Tibet AR.

Appearance

The snow leopard is a large cat with a remarkably long and hairy tail. Males are larger than females. The coat pattern is pale fur scattered with black spots and rings or broken circles. The base pale color varies from light gray to very light brown and is the distinc-

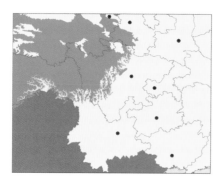

tive feature of the species, besides the smaller body size when compared with leopards. The snow leopard's belly is white, the ears are round and small, and the tail is thick and about the same length as the body. The legs of snow leopards look disproportionably short for a large carnivore.

Habitat and Ecology

Snow leopards are associated with alpine regions in Asia. They use the steep terrain, normally broken by cliffs, ridges, and rocky outcrops, as shelter for denning and raising young; they hunt in alpine meadow and shrubland but usually avoid forests, although snow leopards are occasionally observed in conifer forests and alpine shrublands. They normally occur at elevations between 3,300 and 5,000 m in southwest China but have been detected at higher elevations. In western Sichuan and southern Qinghai Provinces, snow leopards may overlap with leopards, which typically occupy forests at lower elevation. The snow leopard's prin-

cipal prey in this region are blue sheep (*Pseudois nayaur*), as well as smaller prey such as marmots (*Marmota* spp.), pikas (*Ochotona* spp.), and rodents. Increasingly, there are reports of snow leopards killing domestic sheep, goats and yak calves. Snow leopards mate in January and February, and females give birth around May, with a litter size of one to three cubs.

Scat

Snow leopard scats are often segmented with blunt ends, with the exception of a few with pointed ends for the last pellets. Fresh scats have a smooth surface and a diameter of about 1.5–2.5 cm (the largest cat feces in this area). The composition of snow leopard scat is fur and bone. Sometimes the scats may be indistinct from those of wolves when the two species are sympatric.

Tracks

Snow leopards leave one of the largest carnivore tracks (6.5–9.0 cm across) with a typical round cat track shape.

Old scats

Four toes are registered on the track with no claw print (the claws can be seen on slippery surfaces). The toe pads are trapezoid in shape and relatively big; they are bilobate on the front edge and have three lobes on the rear edge.

Other Sign

It is easier to find snow leopard tracks and scat on wildlife trails that follow the boundary between the grassland and rocky ridges and cliffs. Such trails also show evidence of snow leopard scratches. The scratches are typically shallow pits (around 20 × 16 cm) created by the hind feet, sometimes with claw marks and normally with a small pile of dirt and earth on the edge. Such scratches are frequently accompanied by scat.

Snow leopards hunt for large prey (e.g., blue sheep) roughly once per week. They may hide fresh carcasses in sheltered sites of caves or under cliffs and return to consume the remains during the following days. Such kills are frequently found within their home range and may be widely consumed by raptors and other carnivores (e.g., wolves, foxes, bears, badgers).

Fresh scats

Fresh track

Scratching pit

Fresh track

Fresh kill

Tiger

Panthera tigris // Hu

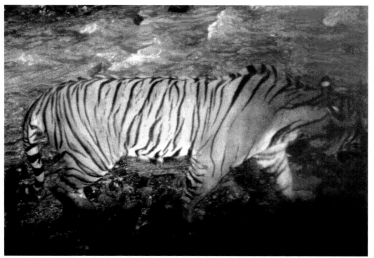

Adult

IUCN: EN
CITES: I
CPS: I
CRL: two subspecies
within this region
(*P. t. corbetti* and
P. t. tigris) CR

BL: 140–280 cm
TL: 91–110 cm
WT: 90–306 kg

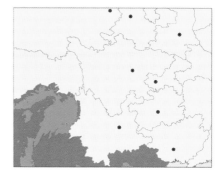

Distribution

Tigers occur in relatively disjunct populations of South Asia, Southeast Asia, far eastern Russia, and scattered sites in China. The historical distribution in China was across the northern and southern regions, but the current distribution is reported only along the boundaries with Russia, India, Myanmar, and probably Laos. In the region in this guide, tigers have been reported in southern Yunnan Province along the boundary between China and Laos (i.e., Indochinese tiger, *P. t. corbetti*) and the boundary between southeast Tibet AR and India (i.e., Bengal tiger, *P. t. tigris*). The population in southern Yunnan Province is probably extinct, with the last individual reported in 2009.

Appearance

The tiger is the world's largest cat with black stripes, so it is easy to distinguish at a glance. Tigers are robustly built with powerful legs, a big head, and a long tail. Their pelage base color is reddish to orange yellow or red brown, except their ventral coat is white. Narrow black stripes stretch from the back to the sides, extending into the ventral coat. There is often a white coloration above the eyes. The tail is thick and longer than half of the head and body length, with black rings.

Habitat and Ecology

Tigers occur in many habitat types, across a broad elevation gradient (sea level to 3,000 m in the Himalayas). Within our focal region all sites are subtropical forest. Tigers prey chiefly on hooved mammals. Their preferred prey species in similar forests are wild boars or deer, which weigh between 10 and 100 kg, but they can kill much larger animals such as gaur (up to 750 kg). Very little is known about tigers' life history in southern China. Studies from other tropical areas indicate their mating season is January and February, with a gestation of 90–105 days and a litter size of two or three cubs.

Scat

The large scats are often segmented, narrow at each end, but not pointed. The diameter of a segment can be more than 3 cm. Tiger scats are composed mostly of undigested animal hairs and bones.

Tracks

The tiger has a large-sized cat track, with adult tracks 8.5–12.5 cm across. The front print is asymmetric. Four toes are registered on the track, and the third one is the largest. The prints from the plantar pads on the foot are trapezoid and relatively large, with two lobes on the front edge and three lobes on the rear edge.

Track

SMALL CATS // XIAO MAO

General Description

Small cats (body weight typically <10 kg) are within the family Felidae and are active and efficient predators living across all habitats and regions of southwest China. Multiple species of small cats may co-occur within this region, such as leopard cats (*Prionailurus bengalensis*) and marbled cats (*Pardofelis marmorata*) in the subtropical forests in western Yunnan Province and southeast Tibet AR and Chinese mountain cats (*Felis bieti*) and Pallas's cats (*Otocolobus manul*) in the alpine grassland in northwest Sichuan and Qinghai Provinces. The tracks and signs (e.g., scats and prey carcasses) of sympatric small cats are usually indistinguishable, but they are readily identifiable from other families of carnivores. Here we present the typical tracks and sign of small cats using the leopard cat as a model.

Tracks

Small cats within the region leave typical round felid tracks with a size comparable to that of house cats (2.5–3.5 cm across for the leopard cat). The front footprint is asymmetric, with four toes registered on the track. The plantar pads are trapezoid and relatively large; they are bilobate on the front edge and have three lobes on the rear edge.

Scat

In general, small cats are highly adaptive to their local environment and prey on a broad range of vertebrate animals, including rodents, pikas, hares, birds, and sometimes reptiles and amphibians. Their scats are usually 1–2

Scat of leopard cat

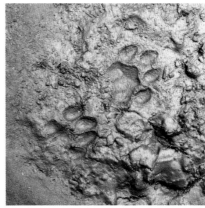

Tracks of leopard cat

cm in diameter, and the shape and color may vary depending on diet. The scats are typically segmented, and the end part is thinner and normally has a needle-shaped tail, which is usually composed of undigested hair, grass, or plant fibers. Their scats are frequently found on elevated surfaces such as logs or stones on a trail. Multiple individuals, sometimes multiple species of sympatric carnivores, may share a latrine. Fresh scats are typically found within scratching pits with disturbed soil, snow, or leaf litter; these scratches are created using the hind feet after defecation.

Asiatic Golden Cat
Catopuma temminckii   Jin Mao
Golden Cat, Temminck's Cat

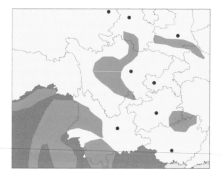

Adult with common golden pelage

IUCN: NT
CITES: I
CPS: II
CRL: CR (evaluated
 as *Pardofelis
 temminckii*)

BL: 71–105 cm
TL: 40–50 cm
WT: 9–16 kg

Distribution

The Asiatic golden cat distribution is from southwest China extending into the Himalayan foothills into Southeast Asia. Within our guide it is found in the mountains of western Sichuan, southern Gansu, and southern Shaanxi Provinces, as well as parts of Yunnan Province and

southeast Tibet AR. It may occur in the northern Guangxi AR.

Appearance

The Asiatic golden cat is a medium-sized felid with a relatively large head, long tail, and stocky body. It has highly variable coat patterns, with the uniform "golden" pelage and the rosette-spotted pelage being the most common. The golden pelage is dark brown to reddish gold on the back and neck, with paler color on the body; the head has sharply contrasted light and dark stripes on the cheeks and forehead. Blurry dark brown spots appear on the belly and the four legs. The second coat color pattern is a spotted, light gray pelage with conspicuous rosette spots (black or dark brown edges with a light brown or yellowish interior) on the flanks and shoulders, black bars on the tail, and dark dorsal stripes on the back and legs. A dark form is also occasionally reported, on which the base color of the pelage is replaced with black or dark gray and there are no conspicuous spots on the body. Different pelage patterns may occur within the same population. The tail is longer than half of the head and body length. Tails are typically curved upward near the tip; the upside of the tail tip is black, and the underside is conspicuously white.

Habitat and Ecology

The Asiatic golden cat habitat ranges from tropical and subtropical evergreen forest to mixed and dry deciduous forest. Asiatic golden cats prefer dense vegetation cover and are seldom observed in open habitat. This cat is solitary and nocturnal and feeds on small-sized vertebrates such as small mammals, rabbits, small deer, birds, and lizards. Pheasants (e.g., blood pheasant and Temminck's tragopan) are a large part of its diet in northern Sichuan. No reports on breeding season exist, and females give birth to one or two kittens.

Scat

Asiatic golden cat scats are one of the largest carnivore scats found in forests in southwest China (diameter of 1.5–2.5 cm). The scat typically progresses from thick to thin, with the thick part normally bent and possibly segmented. Undigested animal remains such as hair, feathers, and bone and grass are normally found in their scats. Fresh scats have a moist, smooth surface and strong odor, and old, dry scats are loose and white in color. The scats are normally found at clear

and open sites along trails, a possible territorial marker.

Tracks

The Asiatic golden cat has a symmetric front track (4.5–5.0 × 5.0–5.5 cm), with four toes registered. The track has a wide plantar pad, which is trapezoid shaped and has three lobes on the rear edge. The hind footprint is smaller than the front footprint.

Spotted form

Hind foot track

Scat

Old scat

Chinese Mountain Cat

Felis bieti // Huang Mo Mao

Chinese Desert Cat, Grass Cat, Chinese Steppe Cat

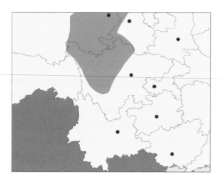

Front view of adult

IUCN: VU
CITES: II
CPS: II
CRL: CR

BL: 60–85 cm
TL: 29–35 cm
WT: 5.5–9 kg

Distribution

The Chinese mountain cat is the only felid species endemic to China. It was formerly recognized as a subspecies of the Asian wildcat (*F. silvestris ornata*). Chinese mountain cats are found only along the eastern edge of the Qinghai-Tibetan Plateau, including eastern Qinghai, northwest Sichuan, and southwest Gansu Provinces. The range shown is increased from that indicated by IUCN on the basis of our local knowledge.

Appearance

The Chinese mountain cat has a uniform base coat of yellow-gray to dark brown color, and the lower lip, chin, and belly are white. Some individuals have indistinct darker stripes on the sides and legs. Two indistinct brownish stripes are seen on the cheeks. The tail is shorter than half of the head and body length, with several black rings and black tip. The ears are triangular and relatively long, with short black tufts on the tip. The winter coat is normally grayer and thicker than the summer coat.

Habitat and Ecology

As the species is rare and sparsely distributed, little is known about the life history of Chinese mountain cats. They are typically observed in bush, steppe, and meadow habitats on the alpine and subalpine elevations from 2,500 to 5,000 m. They feed on small mammals, including rodents, pikas, and hares, and pheasants.

They are solitary and nocturnal. The cats breed from January to March, and the female gives birth to two kittens per litter. Chinese mountain cats have been widely hunted by locals for their fur for clothing materials. Their wild populations may also be threatened by the poisoning of pikas by local herders.

Scat and Tracks

No specific information is available for this species.

Flank view of adult

Leopard Cat

Prionailurus bengalensis // Bao Mao

Adult with summer coat

IUCN: LC
CITES: II
CPS: NL
CRL: VU

BL: 36–66 cm
TL: 20–37 cm
WT: 1.5–5 kg

Distribution

The leopard cat is widely distributed in central Asia, Southeast Asia, South Asia, Far East Russia, and the Korean Peninsula and across China except for the northern and western arid regions. It is found in the entire forest region in southwest China.

Appearance

The leopard cat is the size of a domestic cat. The pelage color in this area is yellowish to light brown on the head, back, body sides, and tail, with a white underside. The whole animal is covered with spots or stripes, including several white and black vertical stripes that go from the nose between the eyes to the head and end at the back of the neck. Variably sized and shaped spots cover the whole body, including the tail. The winter coat is much thicker, with darker spots than the summer coat. The body spots of southern populations are more conspicuous, with clear margins. The tail has numerous black bars on the upper side and is about half the length of the head and body.

Habitat and Ecology

Leopard cats can occur in all forests in southwest China. They occasionally use shrublands and plantations but not grassland habitat. Leopard cats feed on small vertebrates like rodents, birds, rep-

Adult with winter coat

tiles, amphibians, fish, and sometimes carrion. Plant items are also frequently found in their diet, including grass and berries. No specific breeding time is reported; the litter size is two or three kits.

Scat

The scats of leopard cats have typical small-cat scat size (diameter of 1–2 cm), shape, and color depending on the diet (see small-cat overview). Their scats are frequently found on elevated surfaces such as logs or stones on the trail. Multiple individuals may share a common la-

trine. Fresh scats may be found accompanied by scratching pits with disturbed soil, snow, or leaf litter, as created by their hind feet after defecation.

Tracks

Leopard cats leave typical round cat tracks with a size similar to those of house cats (2.5–3.5 cm across). The front footprint is asymmetric. Four toes register on the track. The plantar pads are trapezoid and relatively big; they are bilobate on the front edge and have three lobes on the rear edge.

Scat

Fresh scat, track, and scratching pit

Latrine

Track

Lynx

Lynx lynx // She Li

Eurasian Lynx

Adult

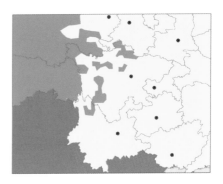

IUCN: LC
CITES: II
CPS: II
CRL: EN

BL: 80–130 cm
TL: 11–25 cm
WT: 18–38 kg

Distribution

The lynx has a wide distribution, occurring along forested mountain ranges in Europe through the boreal forests of Russia, extending into central Asia and the Qinghai-Tibetan Plateau. In the region of this guide, lynx are found in the plateau regions of northwest Yunnan Province, western Sichuan Province, southern Qinghai Province, and eastern Tibet AR.

Appearance

Lynx have a base pelage of gray to gray brown, with varied black or dark brown spots (the spots can be very faint). Lynx pelage in this area is normally pale and not very spotted. The throat and ventral areas are white or light gray. The most distinctive traits of lynx compared with other large cats are the ears and tail, as lynx have long ear tufts and the backs of their ears have light

gray spots. They also have a very short tail with a black tip. A lynx's feet are unusually large with webbing between the toes. The legs of lynx look long compared with those of other cats.

Habitat and Ecology

Lynx are found at alpine and subalpine elevations (3,000–5,000 m) within conifer forest, alpine brushland, meadows, and rocky areas. They feed on small mammals such as rodents, pikas, and marmots and small ungulates. They normally avoid other sympatric large carnivores

Scat

such as snow leopards and wolves. The reported litter size is two or three kits.

Scat

Lynx scat is relatively large (diameter 2.5–3.5 cm) and ropelike, with thin ends and a bend at both ends. Normally, the scat has a needle-shaped end to each segment. Lynx may share latrine sites with sympatric carnivores. Such latrines are typically located at high ridges or cliffs with open views.

Tracks

The lynx has a medium-sized felid print (5.5 × 6.0 cm). The track has four toes that appear relatively large compared with the plantar pad size and are set apart from each other. The print of the toe pads on the front foot is more separated than that of the hind foot. The plantar is triangular shaped with two incisions on the rear edge. The lynx's tracks on snow may show the marks of footpad hair.

Old scat at latrine site

Latrine used by lynx and other carnivores

Front foot track

Marbled Cat

Pardofelis marmorata // Yun Mao

Front view of adult

IUCN: NT
CITES: I
CPS: NL
CRL: CR

BL: 40-66 cm
TL: 36-56 cm
WT: 3-5.5 kg

Distribution

The marbled cat is distributed from Southeast Asia, across the southern Himalayas, into southwest China. There are only a handful reports of marbled cats in China, including the tropical and subtropical forests in southern and western Yunnan Province and the southeast Tibet AR.

Side view of adult

Appearance

The marbled cat resembles a small-sized clouded leopard. The pelage is from yellowish gray to red brown, with dark blotches on the back and body, legs, and tail. The tail is thick and long, almost the same length as the head and body. Unlike many other small cats, marbled cats hold their tail almost horizontal while walking, a distinctive characteristic for this species.

Habitat and Ecology

The marbled cat is a forest-dependent carnivore and is found in moist and mixed deciduous-evergreen tropical forest. This species is poorly known, with few observations in the wild. The diet is likely small mammals, and the marbled cat has been observed hunting arboreal squirrels.

Scat and Tracks

No specific information is available for this species.

Pallas's Cat

Otocolobus manul // Tu Sun

Manul, Steppe Cat

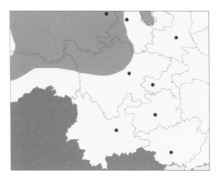

Adult during summer

IUCN: NT
CITES: II
CPS: II
CRL: EN

BL: 45–65cm
TL: 21–35 cm
WT: 2.3–4.5 kg

Distribution

Pallas's cats are distributed from the Middle East through central Asia, Russia, and Mongolia into northern and western China. In our focus area, this cat occurs in the plateau region of southern Qinghai and Gansu Provinces, western Sichuan Province, and part of northern Tibet AR.

Appearance

Pallas's cat is slightly larger than the domestic cat, and the body is stout with obvious short legs and a thick tail. It has dense fur in a grayish color, and the white tip on each hair makes the cat look frosted. Ears are located more laterally on the head than for most fe-

lids. There is a white ring around the eyes, with stripes extending from the eyes to the cheeks. There are small black spots on the forehead. Several faint bands are visible from the back to body sides and front legs, and a bushy tail has black rings and a black tip. The winter coat is grayer and much longer and thicker than the summer coat. The long, coarse hairs on the belly of the winter coat may nearly reach the ground when it is standing.

Habitat and Ecology

Pallas's cats in this region are confined to dry alpine habitat without deep snow cover, including grassland, steppe, and shrubland. Occasionally, the cats are seen on rock crags. Pallas's cats feed on pikas, marmots, small rodents, hares, and birds. They are solitary and nocturnal, hunt by ambush, and are most active at

Adult near den

dawn and dusk. The cats breed in February, and females give birth to a litter of three to six kittens.

Scat and Tracks

No specific information is available for this species.

ELEPHANT // XIANG

Asian Elephant
Elephas maximus // Ya Zhou Xiang
Indian Elephant

Adult male

IUCN: EN
CITES: I
CPS: I
CRL: EN

BL: 5.5–6.5 m
TL: 1.0–1.5 m
WT: 2,500–5,000 kg

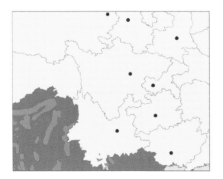

Distribution
Asian elephants are patchily distributed from South Asia through Southeast Asia, with populations isolated within fragmented habitat. In China, Asian elephants were historically found across most of southern China, but the range has dramatically shrunk southward and westward since the 12th century because of habitat loss and hunting. Now they are found only in three prefectures of southern Yunnan Province: Xishuangbanna, Pu'er (previously named Simao), and Lincang.

Appearance
Asian elephants are the largest terrestrial mammals in Asia. They are unmistakable

Herd

animals with a bulky body, huge massive head, distinctive long trunk, and a pair of large ears triangular in shape. There is a single extended lip (upper) at the end of their trunk, which is distinctive from African elephants, whose trunk end has two lips. The Asian elephant's body is covered with thick, wrinkled skin that is normally gray and almost hairless. The body color will turn dark gray to black when wet or yellowish to brownish when covered by dirt or mud. Infants have abundant bristly hairs, and their skin is usually darker. The Asian elephant has a long tail with long black hairs at the tip. Adult males have one pair of long ivory tusks (elongated upper incisors) that are up to 2 m and forward pointing, although most tusks are significantly shorter. Adult females and young also have short tusks, although they rarely protrude from the mouth and are normally not visible.

Habitat and Ecology

Asian elephants inhabit tropical and subtropical habitat types, normally lower than 1,000 m elevation, including forests, shrublands, grasslands, and sometimes plantations. Reports from the eastern Himalayas show that they are occasionally found in montane forests of high elevation up to 3,000 m. They can also be close to human residences within the agriculture-forest landscape. In their remaining range in China they are found in tropical forest remnants and nearby secondary forests. Asian elephants have a broad diet of numerous plants, consisting mostly of grass, palms, and banana stems. When close to people, they also seasonally feed on crops, which causes significant human-elephant conflict. The species is social and lives in herds containing up to 40 individuals. The herd is usually led by a senior adult female, the matriarch, and may consist of other adult females and their offspring of both sexes at varied ages. Adult males may temporarily join these herds or form bachelor herds of multiple individuals (usually <5) or stay solitary. Males will use their strong tusks to fight, defend against threats,

Old, dry dung

Degraded scats

Fresh scats

Left hind leg track

Fresh track

and move obstacles to clear a walking path in the forest. Herds need large areas to forage and move, and individual home ranges may exceed 500 km². In China, home ranges are smaller, but elephants do move widely between forest fragments. Access to surface water is important, and elephants will visit rivers, ponds, and mud wallows almost every day to drink and bathe. They are good swimmers and able to cross large rivers and lakes. Asian elephants have a long life span of 60-70 years in the wild, and the females have the longest gestation among mammals, which takes 18-22 months. The young will stay with the mother within the herd for many years and reach sexual maturity at 10-15 years of age. Asian elephants are important seed dispersers and can translocate seeds over long distances with their dung. Besides habitat loss and fragmentation, poaching is a serious threat, with animals killed mainly for their ivory tusks or because of crop damage.

Scat

Feeding for over 14 hours daily, adult elephants can consume up to 200 kg of vegetation daily and therefore defecate 16-18 times per day, producing a large amount of feces (>100 kg). Fresh scats are large, round piles, each normally 15-20 cm in diameter. The scats contain mainly undigested plant fibers and sometimes fruit seeds. Fresh scats are moist and decompose quickly in the hot, moist climate.

Tracks

Asian elephants leave unmistakable tracks. They have large, round, flat feet, and the tracks are round or oval shaped and large in size (30-45 cm across for adults). Their feet have three toes in the front and one on the side, but the imprints of individual toes are normally indistinct within the footprint. The tracks are normally deeply printed into the soil.

Other Sign

Elephants create obvious feeding sign of broken stems of shrubs, bamboo, and trees. These unmistakable feeding sign are typically found across a large area and in large amounts. The size of the broken stems (up to 12 cm diameter for bamboo) cannot be replicated by other species in this region. Stems can be broken from a height >2 m.

In addition to water, mineral licks are frequently visited. The males may use their tusks to dig earth from a bank or slope and leave deep (up to 50 cm observed) holes. The surface of the vertical bank at these mineral licks can be quite smooth since the elephants lean their bodies and rub against the bank while digging and feeding. Tracks, scats, and feeding signs are also found at and around mineral licks frequented by elephants.

Although mostly silent, elephants can vocalize across a broad frequency range. They use a loud but short trumpet when disturbed; booms at very low

Feeding sign

Mineral lick, vertical bank

Mineral lick

frequency (14–24 Hz, i.e., infrasound that is beyond human hearing) are used to communicate over long distances (maximum of ~10 km), and soft snorts are used within the herd. For feeding elephants, the loud snapping of stems and branches, especially bamboo, is often heard from a distance >100 m.

HOOVED MAMMALS // YOU TI LEI

General Description

Southwest China has 30+ species of hooved mammals that can roughly be grouped together as ungulates. This group contains some primitive lineages such as chevrotains, musk deer, muntjacs, and tufted deer. These lineages either lack antlers (chevrotains, musk deer) or produce short solid antlers on a thick pedicel (muntjacs). For several of these species the canine teeth are extended and protrude outside of the lip (chevrotains, tufted deer, musk deer). The muntjacs are a diverse-lineage small deer found throughout southern Asia, and their distribution extends north into southwest and central China. The taxonomy of the genus *Muntiacus* is still being determined, but the common species within China are identifiable by sight.

One unique attribute of the remaining cervids (and muntjacs) is antlers, a set of bones that annually grow from a pedicel on the skull. These structures are covered with a network of blood vessels and tissue while growing, a velvet that is shed just prior to mating season. The antlers are used in displays and fighting between the males and are shed at the end of the mating season. These bone structures have species-specific patterns of branching and form and are a ready identifier during the right season. Shed antlers in winter and spring are sometimes encountered but are eventually consumed by other mammals. Females do not grow antlers in any of the species in southwest China.

All the cervids and associated species produce a pellet feces, and the collection of pellets is called a fecal pile. If individuals defecate while foraging, the pellets will be more scattered. The standard deer pellet is

Fresh fecal pellets of sambar

Fresh fecal pellets of forest musk deer

Fresh fecal pellets of northern red muntjac

Clumped fecal pellets of northern red muntjac

Fresh fecal pellets of forest musk deer next to older fecal pellets of tufted deer

Dry fecal pellets of sambar

oval, with possibly one end being blunt and dimpled (variations are noted in the text). The pellets may differ in shape, but it is usually the size, combined with the geographic location, that allows species identification. The identification is complicated by the production of similar pellets by bovids in the region. For sympatric species of similar body size, identification of species based on pellets is problematic. One interesting feature is that many species create latrines, where multiple fecal piles are deposited in the same location. We often find several species using the same latrine. The purpose of these latrines is unknown, but they are readily visible in many habitats. Fecal piles are moist and dark when deposited and may shift to a lighter color as they age and dry out. Feces produced from consuming lush vegetation may lose their pellet form and be deposited

Forefoot track of northern red muntjac

as an amorphous mass. For most species the plant material in the feces is well digested, so specific plants or seeds cannot be identified.

All the species in this section are two-hooved mammals, and their tracks are very similar. The hooves are oval shaped and pointed in the front and rounded in the back; the hooves on the hind foot are usually more closely spaced than the hooves of the front foot. Because of their similar appearance, size and geographic location are the primary factors used in track identification. Some species have distinct dewclaws in their track, and some have a more curled tip to the hoof, and these will be noted.

As opposed to the carnivores, many of the cervids are active both day and night, particularly at dawn and dusk, and are therefore more readily encountered.

Williamson's Chevrotain

Tragulus williamsoni // Wei Shi Xiao Xi Lu

Northern Chevrotain, Lesser Mouse Deer

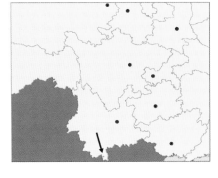

Pair

IUCN: DD
CITES: NL
CPS: I
CRL: CR

BL: 43–50 cm
TL: 6–8 cm
WT: 2.5–4.5 kg

Distribution

Williamson's chevrotain was formerly considered a subspecies of *T. javanicus* and has been recently separated as a distinct species. Williamson's chevrotain are distributed from Yunnan Province, China, into Southeast Asia, although its exact range remains unknown. Within the region of this guide, it is found only in Xishuangbanna Prefecture, southern Yunnan Province.

Appearance

Williamson's chevrotain is the smallest ungulate species within the range of this guide. Its legs are thin, and the back is arched. The forelegs are shorter than the hind legs, so the shoulder appears lower than rump on level ground. The dorsal side is reddish brown, and

the back of the neck is usually darker. The ventral side is white to light yellowish. There are three conspicuous white stripes under the throat down to the chest. Both sexes lack antlers but have one pair of lengthened upper canines, forming long, slightly curved "tusks."

Habitat and Ecology

We know little about the ecology of Williamson's chevrotains. Limited information indicates this species may inhabit tropical evergreen and rainforests. They feed primarily on fallen fruits and young shoots. Williamson's chevrotains are found either solitary or in pairs (male-female pairs or mother-fawn pairs). They are active day and night. Williamson's chevrotains are shy and elusive; when flushed or chased, they have a distinctive jerky and stiff-legged running gait. Habitat loss, bush meat hunting, and trade may be the major threats to this small deer, as with many other tropical ungulates in Asia.

Scat

Williamson's chevrotains produce tiny fecal pellets (<1 cm in length), the smallest among all ungulates within the range. The pellet is typically oval to round.

Tracks

Williamson's chevrotains leave small, hooved prints (around 2 cm), the smallest ungulate track within the range.

Alpine Musk Deer
Moschus chrysogaster // Ma She

Adult female

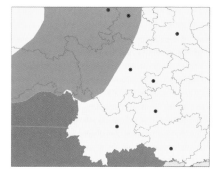

IUCN: EN	BL: 80-90 cm
CITES: II	TL: 4-7 cm
CPS: I	WT: 9-13 kg
CRL: CR	

Gansu, western Sichuan, and northwest Yunnan Provinces.

Distribution

Alpine musk deer are distributed from the southeast to northeast regions of the Qinghai-Tibetan Plateau and parts of the adjacent plateau, with most of their range within China. Within the region of this guide, they are found in western Gansu and Sichuan Provinces, eastern Qinghai Province, northwest Yunnan Province, and eastern Tibet AR. Across its eastern extent, the alpine musk deer may overlap with forest musk deer near the tree line in southern

Appearance

Alpine musk deer are relatively large and stocky compared with other musk deer. Their body is larger (9-13 kg) than the forest musk deer (6-9 kg). Their forelegs are shorter than their hind legs, so the shoulder appears lower than the rump, a common feature shared by all musk deer species. Alpine musk deer have long and sharp hooves, and their dewclaws are well developed. The dorsal side of adults is gray to gray brown, and the ventral side is slightly paler. The lower half of the four legs is pale yellow. Their hairs are stiff and coarse, and the winter coat is denser and darker. There are two pale white to yellow stripes on the side of the throat that join toward the breast; sometimes the two stripes merge as a single broad patch. The throat stripes of alpine musk deer are much paler than that of forest musk deer and are indistinct or almost invisible when observed in the field. Fawns and juveniles are covered with buff spots on the back. On the back of the neck, the hairs are whorled and appear in a banded pattern (normally three to four bands), which is a distinct feature of alpine musk deer compared with similar forest musk deer. The ears are large and long, with many long hairs inside. They have conspicuous dark orange eye rings. Adult males have one pair of long and sharp tusks; the lengthened upper canines are easy to observe.

Habitat and Ecology

Alpine musk deer can inhabit areas above 3,000 m elevation, including alpine meadows, grasslands, shrublands, and the edges of rhododendron, alpine oak, and coniferous forests. In western Sichuan and southwest Gansu Provinces, where they overlap with forest musk deer, the alpine musk deer are normally distributed at higher elevations. They feed primarily on grass and shrubs, and the diet also includes twigs, moss, and lichens. Alpine musk deer are typically solitary or found in mother–fawn pairs. They are crepuscular animals but with notable diurnal activities. Alpine musk deer are shy and agile, and their strong hind legs make them good jumpers. Adults have well-defined home ranges, and the males will mark their territory with scat and musk gland secretions. The male home range may encompass the ranges of multiple females. Mating normally occurs in winter (November–December), and the female will give birth to a single fawn during late spring to early summer (April–June). Numerous predators may prey on the alpine musk deer, including the leopard, wolf, red fox, lynx, and yellow-throated marten. The highly valued musk produced by adult males is widely used in perfume products and for Chinese traditional medicine. The strong demand for their musk has caused severe poaching of all musk deer species. Serious population declines have been reported throughout their range during the last half century.

Scat

Alpine musk deer produce tiny fecal pellets (0.9–1.2 × 0.3–0.5 cm), but they are bigger than those of forest musk deer (0.8–0.9 × 0.3–0.4 cm). The pellet is typically oval, sometimes irregularly shaped, and pointed at one end and dimpled at the other. Fresh fecal pellets are black and normally have a smooth surface. Like forest musk deer, alpine musk deer also have defined latrines along established trails where they defecate regularly, and large piles of scats of varied ages are commonly found at these sites. Shed hairs are

Adult male

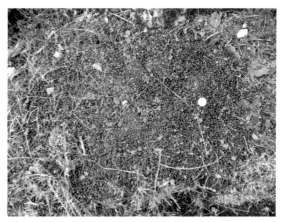

Latrine site

also frequently found on the fecal piles. They also frequently defecate over the scats of other sympatric ungulate species such as tufted deer, Chinese gorals, and Chinese serows. Fresh scats from adult male alpine musk deer may have a strong musk smell.

Tracks

Alpine musk deer tracks can be found in open habitat after a snowfall and usually exhibit distinct prints with well-developed dewclaws. The tracks of the two hooves are elliptical and sharp (6.5–7.5 × 4.5–5.0 cm for the forefoot, 5.0–6.0 × 3.5–4.0 cm for the hind foot).

Forest Musk Deer
Moschus berezovskii // Lin She
Chinese Forest Musk Deer, Dwarf Musk Deer

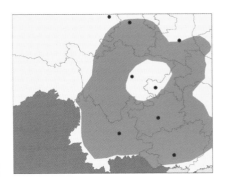

Adult male

IUCN: EN
CITES: II
CPS: I
CRL: CR

BL: 63–80 cm
TL: 4–6 cm
WT: 6–9 kg

Distribution

Forest musk deer are nearly endemic to China, with a small part of their range extending into northern Vietnam and perhaps Lao People's Democratic Republic. They are widely distributed throughout central and southern China (southern Shaanxi, southern Gansu, central Sichuan, Yunnan, and Guizhou Provinces; Guangxi AR; and eastern Chongqing City within the region of this guide). Across its western extent, the forest musk deer may overlap with alpine musk deer near the tree lines in southern Gansu, western Sichuan, and Yunnan Provinces.

Appearance

The forest musk deer is a small ungulate. On level ground its shoulder appears lower than its rump since the forelegs are shorter than the hind legs. Both sexes lack antlers, but the males have one pair of long upper canines that are obvious. The dorsal side of adults is dark olive to dark brown, and the ventral side is yellow orange to light brown. The rump is darker brown to nearly black. There are two pale yellow stripes on the side of throat that join toward the breast. Fawns and juveniles are covered with buff spots on the back. The black-tipped ears are relatively large compared with those of tufted deer and muntjacs, and the inside of the ears is white and covered with long hairs. Forest musk deer have long and sharp hooves, and their dewclaws are well developed, which provides them with the ability to climb on shrubs and lower branches of trees to forage. No other deer would be sighted in trees or shrubs.

Habitat and Ecology

Forest musk deer can inhabit most forests across a broad elevation range from

1,200 to 3,800 m. They are typically solitary or found in pairs. Forest musk deer are shy and agile, as their strong hind legs make them good jumpers. When alarmed, they will quickly jump from the threat and can change direction wildly. Adults have well-defined home ranges, and the male will mark his territory with his scat and musk gland secretions. Forest musk deer are common prey to many predators, including the leopard, Asiatic golden cat, fox, marten, and Asiatic black bear. They maintain a stable home range and trails. Taking advantage of these habits, local poachers usually set foot and head snares along the established trails. Forest musk deer are nervous animals, and once caught in snares, they die quickly. The males have a large gland on the underside of the belly that produces a unique musk scent, which is used to manufacture perfume. The strong, and increasing, demand from the perfume market and Chinese traditional medicine has caused severe poaching of the species. Serious population declines have been reported throughout their range during the last half century.

Scat

Forest musk deer produce tiny rice-shaped fecal pellets (0.8–0.9 × 0.3–0.4 cm). The pellet is typically pointed at one end and dimpled at the other. Fresh fecal pellets normally have a smooth surface, and those from adult males may have a strong musk smell. Forest musk deer have communal latrines, and large piles of scats of varied ages are commonly found. Shed hairs are also frequently found on the fecal piles.

Tracks

Forest musk deer leave unique tracks distinct from those of other small-deer species because of the presence of well-developed dewclaws. The tracks of the two hooves are elliptical and sharp (7 × 4.5 cm for the forefoot, 5.5 × 3.5 cm for the hind foot), and the prints of dewclaws are distinct when imprinted on soft soil or snow.

Other Sign

Shed hair is a common sign of forest musk deer where they forage or rest. The hairs of forest musk deer are easy to recognize since they are very stiff and bristly, and the distal half is wavy.

Hairs

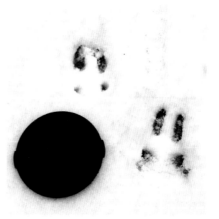

Track

Red Deer

Cervus elaphus // Ma Lu

Wapiti, Shou, Elk

Adult female with summer coat

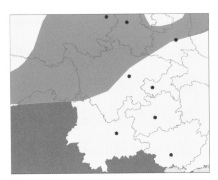

IUCN: LC
CITES: NL
CPS: II
CRL: *C. macneilli,*
  CR; *C. wallichii,*
  EN

BL: 160–265 cm
TL: 10–22 cm
WT: Male, 160–
  240 kg; Female,
  75–170 kg

Distribution

The red deer is the most widespread deer species in the world and has a broad range across the Northern Hemisphere, including Europe, Asia, North America, and part of northern Africa. It has also been widely translocated as a game or farming species. Two subspecies occur within the region of this guide with no overlap in distribution: (1) Sichuan shou, *C. macneilli,* which is found in western Sichuan, southern Gansu, and southern Qinghai Provinces and probably southeast Tibet AR, and (2) Tibetan shou, *C. wallichii,* which is found in southern Tibet AR.

Appearance

The red deer is the second largest deer species (behind the sambar) in south-

west China. The body is stocky with large and broad hooves. Males are much larger than females. The pelage of the red deer varies from summer to winter, from reddish brown in summer to dark grayish brown or dark brown in winter. The summer coat is coarse and short, and the winter coat is thick and long. In both seasons, the ventral side and legs are paler, and there is a dark dorsal stripe in the middle of the back. Adult males grow a long, shaggy mane around the neck during the rutting season in late summer to fall. Both sexes have a large conspicuous yellowish to rusty brown rump patch. The upper margin of the rump patch is dark and may join with the dorsal stripe. The tail is relatively short, and the color is the same as the rump patch. Ears are large and long. Fawns are brighter reddish brown with white or buff spots on the body, which will gradually disappear by the end of their first summer. Adult males have large, stout antlers; compared with the antlers of white-lipped deer, the gap between the first and second tines of red

Adult male

deer is much shorter, and the joints of lower branches are cylindrical rather than flattened. The length of the antlers and the number of tines increase with age before declining in old age. Mature males may have antlers with six to eight tines up to 115 cm in length (from the base to the farthest tip) that weigh 5 kg. Branch joints at the top of the antlers of old males may flatten and form a "cup" or "crown." Females lack antlers.

Habitat and Ecology

Red deer inhabit open deciduous and coniferous forests, alpine shrub, grasslands, and meadows from 2,500 m to 5,000 m elevation within their range in southwest China. They may conduct annual migrations, moving to lower valleys during the winter and back to higher elevations during the summer. Red deer are predominately crepuscular but stay active both day and night. They feed on grass, forbs, moss, lichens, shrubs, and the tender bark and twigs of trees. Red deer typically live in small herds of fewer than 20 individuals, mostly adult females and their offspring. During winter multiple herds may coalesce to form herds of up to 50+ individuals. Adult males are solitary or in small bachelor groups during the nonbreeding season. The rutting season is usually in late summer to fall (late August–October), and the males will compete for females. Rutting males will stop eating, make deep roars, and thrash bushes with their antlers and may lose significant body weight (up to 20%) during this period. Combat between closely matched males is fierce and sometimes fatal. The male antlers are shed during winter and regrow to full length by the next summer. The females give birth to a single fawn in late spring to early summer (June–July). Wolves, bears, and leopards are their primary predators. They have been extensively hunted for their meat and hide, and the large males are hunted for their antlers as trophies, and their velvet (vascular covering during antler growth) is used in traditional Chinese medicine.

Scat

Red deer produce typical large, oval deer pellets (2.0–2.5 × 1.3–1.8 cm), which may be either separated or clumped. When red deer and other large deer species, such as white-lipped deer and sambars, are sympatric, the scats may be indistinguishable between these species.

Tracks

Red deer leave a typical two-hooved track of a large deer. The track size (9.0–11.0 × 8.0–9.0 cm) is one of the largest found within the habitat. Prints in soft soil or snow may show the long dewclaws.

Other Sign

In the fall males frequently rub their body and antlers against trees, which will leave rub marks. Both sexes rub trees to remove their heavy winter coat and thick hairs in spring. However, these rub marks may be indistinguishable from those of other sympatric large ungulates, wild (e.g., white-lipped deer, takin) and domestic (e.g., domestic yaks). During the mating season, the deep roars made by rutting adult males can be heard over a long distance, more common during early dawn and late evening. Loud alarm barks from both sexes are also heard throughout the year.

Sambar

Rusa unicolor   Shui Lu
Indian Sambar

Adult male

IUCN: VU
CITES: NL
CPS: II
CRL: NT (evaluated as *Cervus equinus*)

BL: 180–200 cm
TL: 21–28 cm
WT: 180–260 kg

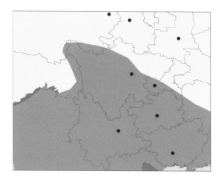

Distribution

Sambars are widely distributed in subtropical and tropical East Asia (including the islands of Hainan and Taiwan), through Southeast Asia, and west along the southern Himalayas to South Asia.

Within the region of this guide, sambars are found in Yunnan and Guizhou Provinces, Guangxi AR, southeast Tibet AR, and most of Sichuan Province.

Appearance

Sambars are a large, stocky deer. Their body pelage is normally dark reddish to brown or black; they are much darker than other sympatric deer species of similar size within this region such as white-lipped deer and red deer (they are called "black deer" by the local people). Legs are usually paler, and their chin is white. Sambars have large, round ears that are whitish inside with dark edges and long hairs at the base. The tail is black with long, brushy hairs and a white underside. Males have long, coarse, and thick hairs around their neck. Unlike many other deer, fawns do not have spots on their coat. Adult males have one pair of three-tined antlers that are large and stout with a maximum length up to 80 cm. Juvenile males (<3 years) may have only one-tine antlers that are nearly straight.

Habitat and Ecology

Sambars inhabit subtropical and tropical forests, but they are highly adaptive to habitats across a broad elevation range from lowland rainforest and swamp forest (<200 m) to alpine conifer forest and shrubland (up to 4,200 m). Sambars frequently visit cropland and plantations, sometimes causing extensive damage and thereby serious human-wildlife conflicts. Sambars are mostly solitary, but small family groups are also frequently observed. They are relatively inactive during the day and forage from dusk through dawn. Sambars are generalist herbivores with a broad diet including grass, ferns, shrubs, young leaves of trees, fruit, etc. Adults can stand on their hind legs to browse young leaves on high branches. Sambars regularly visit mineral licks, especially during the antler-growing season. The mating season is typically from late fall to winter (October–December), when the males will roar and rut. The females will give birth to a single fawn during late spring to early summer (May–July). Sambars are one of the more important prey species for large predators, including tigers, leopards, and dholes, and they are also extensively poached by humans. Severe poaching pressure has caused a dramatic population decline and range reduction in China during the past five decades.

Scat

Large piles of scats are frequently found at their bedding sites and feeding areas. The scats normally include separated or clumped fecal pellets, which are large (2.0–2.5 × 1.3–1.6 cm), oval shaped, and pointed at one end and blunt or dimpled at the other. The

Hind foot track

Adult female

color of scats may vary from brick red to greenish brown and black, depending on diet and the age of the scats.

Tracks

Sambars leave typical two-hooved ungulate tracks, and the size (9.5 × 8 cm for the forefoot, 9 × 6 cm for the hind foot) is usually one of the largest found within their habitat.

Other Sign

Shed antlers from male sambars can occasionally be found. Adult sambars frequently rub their body and antlers (males) against trees or stumps, which will leave rub marks. However, these rub marks may be indistinguishable from those of other large ungulates, such as takins.

Siberian Roe Deer

Capreolus pygargus // Dong Fang Pao

Roe Deer

Adult male

IUCN: LC
CITES: NL
CPS: NL
CRL: NT

BL: 95–140 cm
TL: 2–4 cm
WT: 20–40 kg

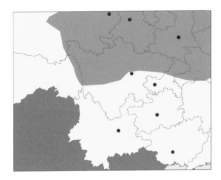

Distribution

Siberian roe deer are widely distributed from Far East Asia through central Asia into eastern Europe. They are a close relative of the European roe deer, *C. capreolus* (Europe to Asia Minor), and were formerly considered a subspecies of *C. capreolus*. Siberian roe deer in China are widely distributed across northern China and are found through the northeast part of the Qinghai-Tibetan Plateau. Within the region of this guide, Siberian roe deer are occasionally found in Gansu and Shaanxi Provinces, as well as western Sichuan and eastern Qinghai Provinces.

Appearance

Siberian roe deer are small, robust deer with a distinctive black band on the muzzle and a white chin. The throat and central breast may have pale patches. Their winter coat is dark gray, and the summer coat is brighter red brown. The ventral side is a paler yellow. The fawn has blurry pale spots on the body that will disappear gradually during the first year. Siberian roe deer have a distinctive white rump with an inconspicuous tail. The rump patch is kidney shaped in males and inverted heart shaped in females. Adult males have one pair of short vertical antlers that are three-tined with a rough surface. The antlers of juvenile males are shorter and lack branches. Females lack antlers.

Habitat and Ecology

Siberian roe deer inhabit forest and forest-meadow mosaics with abundant grass and may avoid areas with dense understory. In their southwest range on the plateau (western Sichuan and southeast Qinghai Provinces), they are found at elevations up to 4,000 m. They are shy and alert animals; when alarmed, their tail will rise, and the white fur of the rump will become erect. They normally flee long distances after detecting humans. Siberian roe deer are primarily solitary in the summer (except for mother-fawn bands) and form mixed groups of up to 20-30 individuals during the winter. They are active mostly during the night and dawn/dusk. Siberian roe deer are polygamous but do not form harems. Their mating season is summer (July–September), and the females give birth, mostly to singles or twins, late the next spring (May–June). Siberian roe deer are important prey species for large carnivores (e.g., leopards and wolves) throughout their range.

Scat

Siberian roe deer produce typical deer fecal pellets, which are typically pointed at one end and dimpled at the other (size 1.3 × 0.7 cm). The fecal pellets can be in piles or scattered, as they defecate while walking.

Tracks

Siberian roe deer leave typical two-toe tracks with pointed tips, as do other small-deer species. The tracks (~5 × 4 cm; larger if the dewclaw is visible) with pointed tips are distinct from those of pigs (wild boars) and goat species (e.g., Chinese gorals), which are more rounded. A trail of tracks, often found on snow, is accompanied by other sign such as browsing on twigs, scats, and resting pits.

Adult female

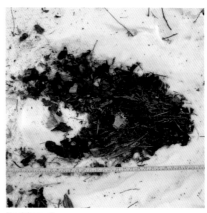

Bedding site

Adult male

Other Sign

Bedding sites for Siberian roe deer can frequently be found on snow during winter. These sites are typically in open areas on the side of vegetation or rocks sheltered from winds and with good visibility. Shed hairs and fresh scats are often found on or around the bedding sites.

Sika Deer

Cervus nippon // Mei Hua Lu

Sichuan Sika, South China Sika, Kopschi Sika

Adult male (left) and juvenile (right)

IUCN: LC
CITES: NL
CPS: I
CRL: Two species: *C. sichuanicus,* and, *C. pseudaxis,* both CR

BL: 105–170 cm
TL: 8–18 cm
WT: Male, 60–150 kg; Female, 45–60 kg

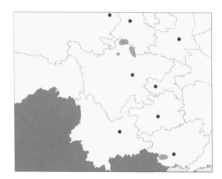

Distribution

Sika deer are widely distributed in East Asia, from Russia to southwest China. They also occur on the islands of Japan and Taiwan. Their current distribution is highly fragmented, and many subspecies and regional populations have been proposed, including two distributed within the range of this guide: (1) Sichuan sika, *C. n. sichuanicus,* occurring in northern Sichuan and southwest Gansu Provinces, and (2) south China sika, *C. n. pseudaxis* (formerly *C. n. kopschi*), occurring in Guangxi AR (the current presence and status of this regional population need further confirmation). Although the distribution of wild populations is highly isolated and restricted, sika deer have been widely farmed in China and have been translocated to many other countries.

Appearance

Sika deer are medium- to large-sized deer with a distinct body pelage. Males are larger than females. The sika deer is one of the few deer species whose adults do not lose the spots that are common in fawns. The overall pelage is bright reddish to reddish brown, with numerous white spots on the back and sides. The ventral side is white. There is a broad black or dark brown dorsal stripe in the middle of back, with one or two rows of white spots closely placed along each side. Both sexes have a small but conspicuous white rump patch, and the upper margin of the rump patch is broad and dark, joining with the dorsal stripe. The tail is short with white margins and underside. Adult males may have a long, shaggy mane around the neck. The winter coat is thicker and darker with less conspicuous spots than the summer coat. Adult males have large antlers with up to five tines

that are slenderer and shorter than the antlers of other large deer species within their range. Females lack antlers.

Habitat and Ecology

Sika deer inhabit deciduous and coniferous forests with dense understory but prefer small open patches inside the forest or along forest edges. They may conduct seasonal vertical migration while living in montane habitats, moving from low valleys during the winter to higher-elevation forests during summer. Sika deer are typically crepuscular but stay active both day and night. They feed on grasses, herbs, shrubs, young leaves, and sometimes fruits. Sika deer are solitary or live in small single-sex herds of fewer than 20 individuals. Larger herds are occasionally observed during winter. The rutting season is in autumn (September–November), when males are territorial and will mark their territories with scraping pits created by hooves and their

Adult females (middle and right) and juvenile

urine. Each male will maintain a small harem of females and defend them against other males. Battles between mature males can be fierce and sometimes fatal. The females give birth to a single fawn, occasionally twins, in the spring (April–June). Males shed their antlers during the winter and regrow them to full length by the next fall. Dholes, bears, and leopards are the primary natural predators of sika deer. Sika deer have been extensively hunted for their meat, hide, and body parts that are used in traditional Chinese medicine, for example, the antler velvet and the penis of males.

Scat

Typical sika deer scats are oval or round fecal pellets (1.5–2.0 × 1.3–1.8 cm), either separated or loosely clumped. Pellets are normally pointed at one end and blunt at the other. Fresh fecal pellets are black, with a smooth and moist surface.

Tracks

Sika deer leave a typical two-hooved deer track (7.0–8.0 × 5.0–7.0 cm), which can be differentiated from those of other deer only on the basis of their size.

Other Sign

During late summer and fall, males frequently rub their antlers and body against trees to remove the velvet and deposit scent, which will leave rub marks. However, these marks may be indistinguishable from those of other sympatric large ungulates, wild (e.g., serows, takins) and domestic (e.g., domestic cattle). Rutting males will create shallow scraping pits with their hooves and urinate into them to mark their territories. Sika deer have various vocalizations, from soft whistles to loud barks.

White-Lipped Deer

Cervus albirostris // Bai Chun Lu

Thorold's Deer

Adult male and females

IUCN: VU
CITES: NL
CPS: I
CRL: EN
BL: 155–210 cm

TL: 10–13 cm
WT: Male, 180–230 kg; Female, 90–160 kg

Distribution

White-lipped deer are endemic to China and occur along the eastern part of the Qinghai-Tibetan Plateau. Within the region of this guide, the species can be found in southern Qinghai and western Sichuan Provinces, as well as northwest Yunnan Province and eastern Tibet AR.

Appearance

The white-lipped deer is one of the larger deer species in southwest China. The body is stocky with relatively short legs, large and broad hooves, and well-developed dewclaws. The pelage of the white-lipped deer is normally reddish brown to grayish brown with coarse hairs; the ventral portions, throat, and

legs are a paler brown. The head and neck are generally darker than the rest of the body. The winter pelage is denser and paler than the summer pelage. White-lipped deer have conspicuous white lips and a white area around the nose. They have large, long ears with white edges near the tip. White-lipped deer have a large rump patch that is generally light to rusty brown with a short tail. Adult males have large, stout antlers that are flattened at branch joints, a distinct characteristic when they are compared with the cylindrical antlers of red deer. Another characteristic differentiating them from those of red deer is the second tine is much farther apart from the first tine (as counted from the base). Mature males may have antlers with a maximum length (from the base to the farthest tip) of up to 140+ cm and up to eight to nine tines. All tines of white-lipped deer lie in roughly the same plane, a third distinctive feature from red deer. Fawns

Mixed herd, adult males and females, and juvenilles

have light spots on the body, which gradually disappear during the first 2-3 months.

Habitat and Ecology

White-lipped deer inhabit coniferous forests, as well as alpine shrub, grasslands, and meadows from 3,500 to 5,100 m elevation. They are also found in rocky montane areas above the tree line. White-lipped deer prefer more open habitats than other large deer species within the range (e.g., red deer, sambars, and sika deer). They are agile climbers in steep, complex terrain and can rapidly flee from large predators (primarily wolves and snow leopards). White-lipped deer are crepuscular animals with some diurnal activity. They typically live in small herds of fewer than 20 individuals, whereas large herds of 100+ (up to 200-300) animals have been reported. The rutting season is usually from late September to November, and the males will compete with each other for females. Each successful male will maintain a small harem of females during the rut. The females give birth to a single fawn before summer (May-June). Out of the mating season, adult males and females are mostly separated. The antlers of males are shed during early spring and regrow to full length by late summer (late August-September). Male white-lipped deer were extensively hunted as trophies, and their velvet (vascular covering during antler growth) is used in traditional Chinese medicine.

Scat

White-lipped deer produce typical deer scats comprising large, oval fecal pellets (2.0-2.5 × 1.4-1.7 cm), which may be separated or clumped. Pellets are normally pointed at one end and blunt or dimpled at the other. The color of scats may vary from greenish brown to black, depending on diet and the age of the scats. When white-lipped deer are sympatric with red deer, the scats of the two species may be indistinguishable.

Tracks

White-lipped deer leave a typical two-hooved track of a large deer, and the size (9.5-10.0 × 8.0-8.5 cm) is usually one of the largest found within the habitat. Prints in soft soil or snow clearly show the long dewclaws.

Other Sign

The male's antlers are shed annually during early spring (around March) and persist on the ground for long periods. During the mating season, males frequently rub their body and antlers

Adult males sparring

against trees or stumps, which will leave rub marks. However, these rub marks may be indistinguishable from those of other sympatric large wild (e.g., red deer) and domestic (e.g., domestic yaks) species.

During the mating season, the growling vocalization made by rutting adult males can be heard over a long distance. Loud alarm barks from both sexes are also occasionally heard throughout the year.

Tufted Deer

Elaphodus cephalophus // Mao Guan Lu

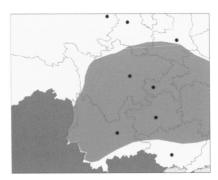

Adult

IUCN: NT	BL: 85–170 cm
CITES: NL	TL: 7–15 cm
CPS: NL	WT: 15–30 kg
CRL: VU	

Distribution

Tufted deer are nearly endemic to China, with a few historical reports from northeast Myanmar. They are widely distributed from southwest to southeast China, and within our region they are found in southern Shaanxi and Gansu Provinces and most of Sichuan, Yunnan, and Guizhou Provinces, as well as northern Guangxi AR and Chongqing City.

Appearance

The tufted deer is a small deer with an overall black to dark brown color. The legs are darker than the body, and the head and neck might be paler in color. There is a distinct tuft of thick black hairs on the top of the head. Its ears are broad and round, with a white edge along the top. The ears show a conspicuous black-and-white pattern that is distinctive from other sympatric ungulates (e.g., muntjacs and musk deer).

The underside of the tail is bright white. When disturbed, the tufted deer will run or jump, with the tail waving rapidly up and down, so the bright white of the tail underside and hind is conspicuous. Males have two small antlers with very short pedicels that are often hidden within the tuft (antler tip normally

Adult male

less than 2 cm above tuft). Males also have long upper canines that are visible outside the mouth.

Habitat and Ecology

Tufted deer are found in montane forests across a wide elevation range up to 4,000 m. Their habitat includes forest, shrub, and early successional vegetation. Although widely distributed, tufted deer are poorly studied, and little is known about their ecology. The deer are primarily diurnal, and most observed deer are solitary or occasionally in pairs. Tufted deer have a broad diet, including grass, forbs, leaves, bamboo, and mushrooms. They will regularly visit natural or artificial salt licks within their home range. They may conduct seasonal migrations along an elevation gradient and use higher elevations during the summer and lower forests or open shrublands during the winter to access forage and avoid deep snow.

Comparison of fecal pellets of tufted deer (left and middle) and Reeves's muntjac (right)

Scat

The separate pellets are normally oval shaped or slightly triangular shaped, with an indistinct dimple at one end and no point at the other. The size (1.2 × 0.6 cm) and shape of their scats may partially overlap with those from other sympatric ungulates of similar size (e.g., red muntjac and Chinese goral), which makes species identification difficult when based solely on scats.

Tracks

Tufted deer leave a typical cervid track (4–4.5 × 3–3.5 cm) that is not distinct from those of other ungulates of similar body size.

Gongshan Muntjac

Muntiacus gongshanensis // Gong Shan Ji

Adult female

IUCN: DD
CITES: NL
CPS: NL
CRL: CR

BL: 95–105 cm
TL: 9–16 cm
WT: 16–24 kg

Distribution

The Gongshan muntjac is one of the least known ungulate species in China. It has been reported to be distributed in the Gaoligong Mountains, northwest Yunnan Province, and southeast Tibet AR (Mêdog and Bomi Counties), probably extending to northern Myanmar (Kachin State). There are disputes over the taxonomic lineage of this species. Sympatric muntjac species with similar dark pelage occur within this region, and visual identification can be difficult.

Appearance

The Gongshan muntjac is medium-sized. The dorsal side is dark brown; the ventral side and legs are nearly black. The tail is black, but the underside is bright white. Adult males have two simple antlers with short, thick pedicels (base of antler). The front sides of pedicels are covered with black hair. The two pedi-

cels extend as bony ridges and join at the forehead to make a conspicuous V shape. Their antlers are 7–8 cm in length and smaller than those of the northern red muntjac. Females do not have antlers, but the V-shaped dark lines are also conspicuous on the forehead. Both sexes lack forehead tuft, which is distinct for the black muntjac.

Habitat and Ecology

The Gongshan muntjac has been rarely studied, and little is known about its natural history and ecology. A handful of specimens and camera-trapping records suggest this species may inhabit a broad elevation range from 900 to 3,000 m across various habitats, including subtropical, temperate, conifer, and alpine broadleaf forest and alpine scrubland.

Scat and Tracks

There is no available information on the track, scat, and other sign of the Gongshan muntjac. They may be indistinguishable from those of other sympatric muntjac species (e.g., northern red muntjac).

Northern Red Muntjac

Muntiacus vaginalis // Chi Ji

Red Muntjac, Indian Muntjac, Barking Deer

Adult male

IUCN: LC
CITES: NL
CPS: NL
CRL: NT

BL: 98-120 cm
TL: 17-20 cm
WT: 17-40 kg

Distribution

The northern red muntjac is the most widespread muntjac species in Asia and is widely distributed from South Asia through northern Southeast Asia into southern China. Within the region of this guide, it is historically found in Guizhou and Yunnan Provinces, Guangxi AR, and southern Tibet AR, although the current distribution is unclear and probably reduced to a handful of areas.

Appearance

Northern red muntjacs are a large muntjac species with a bright reddish to red brown pelage. Their dorsal side varies from dark reddish to rufous, and the ventral side is paler gray to whitish. The underpart of the tail is bright white. They

have large preorbital glands that are conspicuous on the side of the head, especially for the males. Adult males have two simple antlers with sharp, curved tips and a small spike near the base. The male's pedicels (base of the antlers) are much longer and wider than those of Reeves's muntjac. The front sides of pedicels are normally covered with black fur. The two pedicels extend as bony ridges and join at the forehead to make a conspicuous V shape. Females lack antlers but have a hair tuft that can be indistinct. Males shed their antlers annually during the nonbreeding season. Fawns and juveniles have white spots that will disappear with maturity.

Habitat and Ecology

Northern red muntjacs inhabit a variety of temperate forests, from lowland rainforest to montane deciduous forest, and are sometimes found in open habitat or close to tree plantations. Their diet consists of twigs, buds, young

Adult female

leaves, fallen fruits, and seeds, and they are an important seed disperser. Northern red muntjacs are typically solitary except for mated pairs, and mother-offspring units are occasionally observed. They are active day and night and breed throughout the year, although a seasonal birth peak has been reported within some populations. Females normally give birth to a single fawn. Northern red muntjacs are important prey for large carnivores. They are also heavily hunted by locals for meat; their overall populations are decreasing, and many local populations have been eliminated.

Scat

The long oval-shaped pellets (1 × 0.5 cm) are typically blunt or slightly pointed at one end, and there is one shallow dimple at the other end. The color and shape of pellets may vary broadly depending on diet. Fruit seeds are commonly found in scats.

Tracks

Northern red muntjacs' tracks are about the largest (3 × 2.5 cm) muntjac tracks. Their tracks cannot be distinguished from those of other muntjac species except for the larger size.

Other Sign

Adults of both sexes make loud barking calls that can be heard at a distance.

Reeves's Muntjac

Muntiacus reevesi // Xiao Ji

Chinese Muntjac, Yellow Muntjac

Adult male

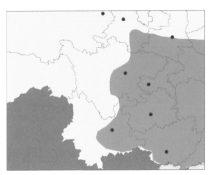

IUCN: LC
CITES: NL
CPS: NL
CRL: VU

BL: 64–90 cm
TL: 8–13 cm
WT: 11–16 kg

Distribution

Reeves's muntjac is endemic to China and widely distributed from southwest to southeast China and the island of Taiwan. Within the region of this guide, Reeves's muntjac is found in all the eastern provinces (Guizhou, Chongqing City, and Guangxi AR) as well as southern Shaanxi, southeast Gansu, and eastern Sichuan and Yunnan Provinces.

Appearance

Reeves's muntjac is a small yellowish deer, whose dorsal side varies from yellow to brownish and whose ventral side is much paler. Its winter coat is darker, longer, and thicker than its summer coat. The tail is light brownish with a bright white underside. Males have two small antlers with sharp tips and a small spike near the base. The male's pedicels

(base of the antlers) are much shorter than those of the red muntjac. The front of the pedicels is covered with black hair extending down to the forehead and making a conspicuous V shape. Females lack antlers and have a black diamond-shaped patch in the center of the forehead. Fawns and juveniles have indistinct pale spots on the body that disappear gradually with age.

Habitat and Ecology

Reeves's muntjacs live in temperate, subtropical, and tropical forests. They can also utilize planted conifer forest and shrubby habitat. Although they can be found in montane habitat, the species typically lives at lower elevations, <2,700 m. They are solitary or found in pairs. An adult home range is less than 100 ha, and individuals have high site fidelity. There is no clear seasonality to their reproduction, and females are sexually mature within 1 year of age. They will regularly visit natural or artificial mineral licks within their home range.

Female with fawn

Individuals are highly vigilant and, when disturbed, will flee immediately, with the tail waving quickly up and down to present the bright white rear and tail underside. Reeves's muntjacs are curious animals and will stop at a distance to observe after fleeing.

Scat

Reeves's muntjacs produce typical ungulate fecal pellets, the smallest (0.8 × 0.5 cm) except for those of musk deer and chevrotains. The pellets are typically pointed at one end, and the flank has one or more dimples. The cross profile is usually triangular, whereas musk deer pellets have a round or oval profile.

Tracks

The tracks (2.5 × 2 cm) are among the smallest of the hooved animals within the range, except for musk deer and chevrotains. These tracks are distinct from those of musk deer in the absence of prints from the dewclaws (i.e., musk deer show their dewclaws; see forest musk deer).

Other Sign

Adult Reeves's muntjacs make a loud barking call during both day and night that can be heard at a long distance.

Argali

Ovis ammon // Pan Yang
Wild Sheep, Mountain Sheep

Adult males

IUCN: NT
CITES: *O. a. hodgsoni*, I; all other subspecies, II
CPS: II

CRL: NT (evaluated as Tibetan argali, *Ovis hodgsoni*)
BL: 180–205 cm
TL: 10–17 cm
WT: 45–180 kg

Distribution

Argali are distributed widely from central Asia to much of the Qinghai-Tibetan Plateau and Mongolian Plateau, extending to southern Siberia. Historical records include a larger extension east to northern China (Shanxi Province and the adjacent mountainous areas). Numerous subspecies have been recognized across their range. Within the region of this guide, there is only one subspecies of argali (*O. a. hodgsoni*, sometimes referred to as Tibetan argali, *O. hodgsoni*), which occurs in Qinghai, southwest Gansu, and western Sichuan Provinces, as well as Tibet AR.

Appearance

Argali are stocky sheep-like ungulates, with adults having massive curved horns

and a very short tail. They are the largest species of wild sheep in the world. Adult males (95–180 kg, maximum over 300 kg) are much larger than females (45–100 kg). Their dorsal pelage is grayish brown to yellowish brown, whereas the ventral pelage and rump are white to light gray. Adults, especially the males, have dark stripes on the flanks and the front edges of the four legs. Adult males have a conspicuous white neck ruff with long hairs down to the breast. Their winter coat is much thicker than the summer coat. Both sexes have one pair of horns, but the males' horns are much stouter and longer, up to 190 cm in length and weighing up to 23 kg. The males' horns have heavy annuli and a broad base and curve down and then forward, slightly flaring outward, for more than 360°. Females' horns are much smaller, generally less than 50 cm in length, and just slightly curved.

Habitat and Ecology

Argali are typically found in open or steep terrain at high elevation from

Female

3,000 to 5,800 m. They primarily inhabit alpine meadow, grassland, rocky outcrops, desert, and semiarid habitats. During winter they may migrate to lower elevations. Their diet includes grasses, sedges, forbs, and lichens, and they are active throughout the day. They are typically observed in herds of 2–100 individuals, but larger herds of up to 200 individuals are occasionally observed. Males may form bachelor groups during nonbreeding seasons. Argali are agile climbers in steep and rocky terrain, and their pelage results in their being well camouflaged. Males tend to inhabit higher elevations than females. To escape from predators, that is, snow leopards and wolves, females with lambs may occupy steeper areas. The rutting season is winter, from October to January, during which mature males will spar with each other, using head butting to establish their dominance.

Females give birth to a single lamb in spring (March–April), during which the mother and lamb may separate from the herd for a couple of days. Poaching, primarily for trophy and meat, is considered their top threat. Competition with and disease transmission from livestock (primarily domestic yak and sheep) are another substantial threat. As a result, many populations have been extirpated during the past half century in southeast Qinghai, western Sichuan, and southern Gansu Provinces.

Scat

Argali scats are typical sheep-like fecal pellets but of larger size. Separated fecal pellets are typically oval, pointed at one end and blunt at the other. Pellets are sometimes irregularly shaped or loosely clumped. Their pellets are often found spread across a large open area as a result of individuals moving as they forage.

Tracks

Argali primarily occur in steep rocky and open alpine habitats, so few tracks are observed outside of the winter season. Tracks can be found after a snowfall.

Other Sign

During the winter mating season, the head butting of rutting males can be heard over a long distance.

Fecal pellets

Habitat

Blue Sheep

Pseudois nayaur // Yan Yang

Bharal

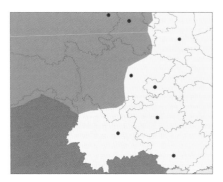

Front and back view of adult male

IUCN: LC
CITES: III
CPS: II
CRL: LC

BL: 110–165 cm
TL: 10–20 cm
WT: 35–80 kg

Distribution

Blue sheep are mainly distributed on the Qinghai-Tibet Plateau and in the surrounding mountains. Within the region of this guide, they occur in Qinghai and Gansu Provinces, western Sichuan and northwest Yunnan Provinces, and the Tibet AR.

Appearance

Blue sheep are stocky goatlike ungulates with distinctive horns and a short, black tail. Males (50–80 kg) are much larger than females (35–45 kg), and their neck girth is stronger and larger. Their dorsal side is brownish gray to blue gray; the ventral side and rump are white to light gray. The inner sides of the legs are whitish, whereas the front edges of the four legs have conspicuous black stripes. Adult males have a black chest and forehead and one black stripe along each flank. Lambs lack most of the adults' black markings. Their winter coat is much thicker than that in summer. Both sexes have one pair of smooth horns, but the males' horns are much stouter and longer, up to 90 cm, curving backward, then flaring outward with a twist. Blue sheep have well-developed dewclaws.

Habitat and Ecology

Blue sheep are usually found in open or steep terrain at high elevation from 3,000 to 6,500 m. They primarily inhabit alpine meadow, grassland, and rocky outcrops and debris fields but occasionally visit alpine shrubland and rhododendron and conifer forest as low as 2,500 m elevation. Blue sheep feed on grass, forbs, and lichens and are active throughout the day. They are social herbivores and are typically observed in groups of 10–40, but large herds of up to 300 individuals are recorded. Males may form bachelor groups during nonbreeding seasons. Blue sheep are agile climbers in steep and rocky terrain, and their pelage results in their being well camouflaged. Their habitat contains several

Side view of juvenile male

Adult female and lamb

Fecal pellets

large predators (i.e., snow leopards and wolves), and blue sheep are the primary prey for snow leopards. Blue sheep mate during the winter, and females give birth to one offspring in early summer. Although the overall population of blue sheep is considered stable, local populations in southwest China are subject to high poaching and poisoning rates for their meat, fur, and skull.

Scat

Blue sheep scats are typical goat-like fecal pellets, either separated or loosely clumped. Separated fecal pel-lets (1.4–1.6 × 1.1–1.4 cm) are typically round, with a distinct dimple at one end and a point at the other. Pellets are sometimes oval to irregularly shaped. Their pellets are often found spreading across a large open area, as individuals defecate while foraging.

Tracks

Blue sheep primarily occur in steep rocky areas and alpine meadows, so few tracks are observed outside of the winter season. Tracks can be found after snowfalls and, in most cases, are easy to identify since few other ungulates of similar size co-occur at their elevation.

Chinese Goral

Naemorhedus griseus // Zhong Hua Ban Ling
Grey Long-Tailed Goral, Sichuan Goral

Dark pelage form

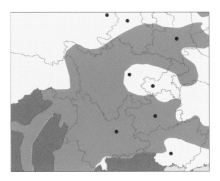

IUCN: VU	BL: 88–118 cm
CITES: I	TL: 11–20 cm
CPS: NL	WT: 22–32 kg
CRL: VU	

Distribution

Chinese gorals are widely distributed in central and southwest China, and their range extends into Southeast Asia. Within the region of this guide, they occur in all provinces except for Qinghai Province.

Appearance

Chinese gorals are goatlike ungulates with a long, bushy black tail. They have a pair of sharp, slender horns that are slightly curved and point backward. They have multiple coat colors, which vary from pale gray to brown and almost black. Most individuals have a darker stripe in the middle of the back. The lower part of the leg is paler than the body. They have an obvious white or yellowish throat contrasting with the rest of the body.

Habitat and Ecology

Chinese gorals are usually found in steep terrain, either open or forested, across a broad elevation range from 1,000 to 4,400 m. They are agile rock climbers and are frequently observed on cliffs and mountain ridges. They are active during both day and night and can be seen during the day foraging near or on

Pale gray pelage form

ledges and outcropping. Chinese gorals can be solitary, in pairs, or in small groups of multiple adults. They feed on numerous forbs and grasses, including bamboo. High mortality has been reported in spring (March–May), during which time goral carcasses are frequently found near rivers or creeks.

Scat

Chinese goral scats vary by season from separate fecal pellets to a clumped single mass depending on diet. Separated fecal pellets (1.4 × 1.2 cm) are typically round, with a distinct dimple at one end and point at the other. Their pellets have a smaller size and rounder shape that are distinct from those of the Chinese serow. Pellets are sometimes oval to irregularly shaped and clumped.

Tracks

Most goral activity is usually in rocky areas, and no tracks are left. Tracks (3.5 × 3 cm) can be found on soft ground surface but are similar to those of other ungulates of similar size.

Track

Fecal pellets (top) compared with pellets of a Chinese serow (bottom)

Brownish pelage form

Himalayan Goral

Naemorhedus goral // Xi Ma La Ya Ban Ling

Himalayan Brown Goral, Eastern Gray Goral

Adult male

IUCN: NT
CITES: I
CPS: II
CRL: EN

BL: 82–120 cm
TL: 8–20 cm
WT: 35–42 kg

Distribution

The Himalayan goral has a narrow distribution along the Himalayas including parts of China, Indian, Bhutan, Nepal, and Pakistan. Within the region of this guide, Himalayan gorals occur only in southern Tibet in the region contested by China and India.

Appearance

Himalayan gorals are medium-sized goatlike ungulates. Their pelage is rufous brown to reddish brown with coarse, black-tipped guard hairs. The ventral side is paler, and the lower part of the legs is light brownish to tan. Throat and chin are contrasting white. The winter coat is shaggy with thick underfur. There is a dark dorsal stripe in the middle of back. Old males have a semierect black mane on the back of the neck. The tail is long, and the distal part is black and not bushy. Both sexes have one pair of black slender horns (12–18 cm in length) that are slightly curved and point backward. The lower part of the horn is ringed, and the upper part is smooth and sharp. Horns of females are shorter and slenderer than those of males.

Habitat and Ecology

Himalayan gorals inhabit steep mountainous terrain in the Himalayas, either in open areas or with shrub and forest cover, across a broad elevation range from 1,000 to 4,000 m. They feed primarily on grass, young leaves, twigs, and fruits. Himalayan gorals are nimble rock climbers and are typically found foraging on rough, rocky slopes and grassy ridges or cliffs. They rest and shelter in forested areas or rock overhangs. They are primarily diurnal but may stay active during both day and night. They are solitary or in groups of two to five individuals. Males are territorial. Both sexes reach maturity by 3 years, and the females give birth to single or twin calves. Their natural predators include dholes, leopards, and tigers.

Track and Scat

We have no specific information, but the tracks and scats of Himalayan gorals are similar to those of Chinese gorals.

Red Goral
Naemorhedus baileyi // Chi Ban Ling

Adult

IUCN: VU
CITES: I
CPS: I
CRL: EN

BL: 93–107 cm
TL: 8–10 cm
WT: 20–30 kg

Distribution

Red gorals are distributed in a narrow range along the eastern Himalaya. Within China they can be found in the mountainous area of southeast Tibet AR and northwest Yunnan Province (Gongshan County).

Appearance

Red gorals look similar to Chinese gorals, but the coat color is dark red to warm brown. They have a dark brown to black tail that is relatively short compared with that of Chinese gorals. The legs, especially the lower parts, and the throat are paler than those of Chinse gorals. There is a thin, sometimes unclear, dorsal stripe that is dark brown. The ridge of the nose is darker than the rest of the head. As with other gorals, red gorals have two slender horns that are slightly curved and point backward.

Habitat and Ecology

Red gorals are rare, and the total population is estimated to be less than 10,000 individuals. They are typically found in steep terrain at an elevation from 2,000 to 4,500 m, higher than most other goral species. They prefer primary coniferous forests but are also found in nearby alpine meadows and shrub habitats. They are agile

Fresh fecal pellets

rock climbers and are often observed near cliffs and ledges. Red gorals are mostly diurnal and solitary but are occasionally seen in groups of multiple individuals, probably an adult with offspring. Red gorals may conduct annual vertical migrations, moving up to near the tree line during the summer and back to lower elevation during the winter to avoid deep snow and access green forage.

Scat

Red goral scats vary by diet. Separated fecal pellets are typically oval, sometimes irregularly shaped (1.0 × 0.9 cm), with a distinct point at one end. Fresh scats are often dark brown in color when red gorals consume a large amount of lichens and conifers. Pellets are sometimes clumped when the diet is dry forage.

Tracks

Red goral tracks (3.5 × 3 cm) are similar to those of Chinese gorals, with two elliptical spheres that are blunt in the front and taper in the back. Tracks can be found on soft ground surface but are similar to those of other ungulates of similar size such as muntjacs and tufted deer.

Adult

Chinese Serow

Capricornis milneedwardsii // Zhong Hua Lie Ling
Southwest China Serow

Adult

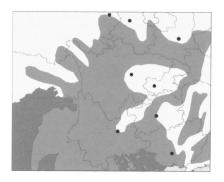

IUCN: NT	BL: 140–170 cm
CITES: I	TL: 11–16 cm
CPS: II	WT: 85–140 kg
CRL: VU	

Distribution

Chinese serows are widely distributed from central China, through the eastern Himalayas, and into Southeast Asia. Within the region of this guide, Chinese serows are found in forest regions of all provinces and ARs.

Appearance

Chinese serows are robust, long-legged, goatlike ungulates and are much larger (3–6 times heavier) than gorals. Their body pelage is mostly black, with contrasting reddish or light brown on the lower part of the legs. The ventral side is often paler. Chinese serows have a remarkable long and shaggy mane on the back of the neck that is normally white or beige. Their coat hairs are quite coarse. The throat of the Chinese serow is often white to light brown, forming

a paler patch. Their ears are long and large, for which they are called mountain donkeys by the local people in southwest China. They have one pair of horns that are similar to those of gorals but much stouter, less curved, and with more circular ridges near the base.

Habitat and Ecology

Chinese serows are found in most forest types across a broad elevation range, from lowland rainforest and karst shrubland (~200 m elevation) to alpine conifer forest up to 4,500 m elevation. Little is known about their natural history and ecology. They may naturally occur at relatively low density compared with other sympatric bovids such as gorals and takins. Chinese serows are typically shy and solitary and have high fidelity to their home range. They will regularly visit the same areas within the home range by following established trails. Animals living along the eastern edge of the Qinghai-Tibet Plateau may conduct seasonal migrations along the elevation gradient. Similar to Chinese gorals, Chinese serows have been reported to experience high mortality in early spring (March–May), during which time their carcasses are frequently found near rivers and streams.

Scat

Chinese serows establish latrines along their routine trails, where they regularly defecate. The scats normally include hundreds of separated fecal pellets, which are large (2.5–3 × 1.2–1.6 cm), oval shaped, pointed at one end, and blunt or sometimes with a dimple at the other.

Tracks

Chinese serows leave typical two-hooved ungulate tracks, and the size (7 × 9 cm) is intermediate to those of gorals and takins. On soft ground, the hoof prints will be more spread out, and the dewclaws might be visible.

Fecal pellets at different ages

Himalayan Serow

Capricornis thar // Xi Ma La Ya Lie Ling

Adult

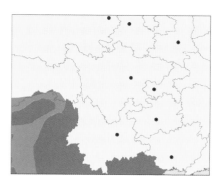

IUCN: NT	BL: 140–170 cm
CITES: I	TL: 11–16 cm
CPS: II	WT: 85–140 kg
CRL: EN	

Distribution

Himalayan serows are distributed in a narrow region along the southern slope of the Himalayas (China, Bangladesh, Bhutan, India, and Nepal), and their range extends south to northeast India and western Myanmar. Within the region of this guide, Himalayan serows are found only in the southern Tibet AR.

Appearance

Himalayan serows are large and robust goatlike ungulates with stocky legs. The overall appearance is similar to that of the Chinese serow. The dorsal side is black with a dark central stripe on the back. The ventral side is paler, and the legs and rump are contrasting reddish to orange. The throat is beige to light brown, and lips are white. Adults have a shaggy mane on the back of the neck that is normally beige to black and much shorter than that of the Chinese serow.

The tail is short and black. Their donkey-like ears are long and large, with white hairs along the inside edge. Himalayan serows have one pair of horns that are similar to those of the Chinese serow, but much stouter and less curved, almost directed straight backward. Their horns are also smoother and less annulated than those of the Chinese serow.

Habitat and Ecology

The basic biology and ecology of Himalayan serows are poorly known. Limited information indicates this species inhabits montane forest up to 3,000 m elevation along the southern slope of the Himalayas. They are often observed in the steep terrain of cliffs and rock outcroppings. Dholes might be the primary predator of the Himalayan serow. Although their population status and trends are unknown, Himalayan serows have been heavily hunted and poached, primarily for their meat.

Scat and Track

No specific information is available for this species, but they are probably similar to those of the Chinese serow.

Przewalski's Gazelle

Procapra przewalskii // Pu Shi Yuan Ling

Adult male

IUCN: EN	BL: 109–160 cm
CITES: NL	TL: 7–12 cm
CPS: I	WT: 17–32 kg
CRL: CR	

Distribution

Przewalski's gazelles were historically distributed across much of the Tibetan Plateau, but their range has dramatically shrunk during the past century. The current range is confined to small areas around Qinghai Lake and portions of Tianjun and Gonghe Counties in Qinghai Province.

Appearance

Przewalski's gazelles are medium-sized gazelles with a short and blunt muzzle. Their pelage is similar to that of their close relative, the Tibetan gazelle, whereas their body size is larger and stockier. They are agile animals with long and thin legs. The dorsal coat is sandy brown to gray brown; the ventral side and throat are whitish. The winter coat is thicker and fluffier than the summer coat, and the color is paler. Both sexes have conspicuous white rump patches, which are easily recognized at a distance. Unlike the heart-shaped rump patch of the Tibetan gazelle, the rump patch of Przewalski's gazelle is divided into two parts by the central dark line. The tail is black and short. Adult males have one pair of heavily annulated horns (around 30 cm in length) that are similar to those of the Tibetan gazelle but thicker and with the ends pointing toward each other instead of being parallel to each other. The females lack horns.

Juvenile female

Habitat and Ecology

Przewalski's gazelles inhabit the open habitats around Qinghai Lake on the plateau around 3,200–3,400 m elevation, including grasslands, dunes, meadows, and semideserts. They are one of the most endangered ungulate species in China, with the total population estimated to be less than 1,500 individuals. In Tianjun County Przewalski's gazelles overlap with the Tibetan gazelle, but they are mostly found at lower, flatter terrain. Przewalski's gazelles are social grazers and are normally found in small herds of 2–20 individuals, although larger herds of over 30 animals are occasionally observed. The gazelles are diurnal, and their diet contains primarily steppe grasses. Adult females and males are separate during the nonbreeding seasons, and mating generally occurs during the winter (November–December). Females will give birth to single offspring (occasionally twins) during early summer (May–July). Wolves are their primary predator, but the conflicts with livestock and the fences established by herdsmen are also major threats.

Scat

The scats are typically oval, sometimes irregularly shaped, fecal pellets (0.5–0.8 × 0.3–0.4 cm). There is a pointed tip at one end of the fecal pellet. The pellets may be clumped, especially during the rainy and growing season (June–August).

Tracks

Tracks are frequently found, especially after summer rain or winter snow. The prints of the two hooves of the forefoot are more separated than those of the hind foot.

Other Sign

Adult males create numerous scratching pits during the rutting season (November–December). These scratching pits are conspicuous, especially on lightly snow covered ground. Scats are generally found around the pits. Bedding sites are sometimes found and may be clumped as family herds stay together; scats are normally found nearby. Carcasses, bones, and body remains of Przewalski's gazelles are frequently found along pasture fences within their habitat as individuals become entangled while crossing the fences. Wolf killing and scavenging sites are also found near fences.

Track

Scratching pits and scats

Fresh fecal pellets

Habitat

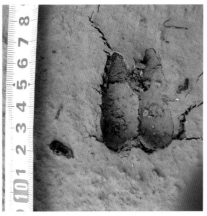
Forefoot print

Resting beds

Tibetan Gazelle

Procapra picticaudata // Zang Yuan Ling

Goa

Adult male with summer coat

IUCN: NT
CITES: NL
CPS: II
CRL: NT

BL: 91–105 cm
TL: 8–10 cm
WT: 13–16 kg

Distribution

Tibetan gazelles are native to the Qinghai-Tibetan Plateau and are widely distributed across the region, with most of their range within China. Within the region of this guide, they occur in western Sichuan, southwest Gansu,

Adult females in winter coats

and most of Qinghai Provinces and northern Tibet AR.

Appearance

Tibetan gazelles are small-sized antelope with a short and blunt muzzle. They are slender and agile animals with long and thin legs. The dorsal coat is light brown to gray brown, and the ventral side is white. The winter coat is thicker and fluffier than the summer coat, and the color is paler. Both sexes have a conspicuous heart-shaped white rump patch, which can be easily recognized at a distance. The tail is black and short. Adult males have one pair of slender, heavily annulated horns (26–32 cm in length), and horns are absent from females. The horns grow upward from the back of the head and curve backward near the tips, and then the tips point upward again. The upper parts of the two horns, including the tips, are almost parallel to each other, which is a distinct characteristic when Tibetan gazelles are compared with the similar Przewalski's gazelle.

Habitat and Ecology

Tibetan gazelles inhabit open habitats on the plateau from 3,000 to 5,750 m elevation, including grasslands, meadows, alpine shrublands, and semideserts. Tibetan gazelles are social grazers and are normally found in small herds of 2–20 individuals, with larger herds of up to 50 animals occasionally observed. Tibetan gazelles are wary animals; they are vigilant to any moving objects on the open terrain. They will quickly flee when alerted but will stop at a distance and turn their head back to inspect. Tibetan gazelles are diurnal animals. They forage primarily on forbs, grasses, sedges, and legumes. Adult females and males are mostly separate during the nonbreeding seasons, and mating generally occurs during the winter (around December). Females will give birth to a single offspring (occasionally twins) the next summer (July–August). Wolves are the primary predator of Tibetan gazelles. Fences established by herdsmen on the rangeland of western China are considered a major threat.

Adult male with winter coat

Scat

The scats are typically round to oval, sometimes irregularly shaped, fecal pellets. They are the smallest in size (0.6–0.8 cm) among all wild ungulates occurring on most of the plateau. Pellets may be clumped, especially during the rainy and growing season (June–August) when Tibetan gazelles feed on alpine forbs.

Tracks

Tracks are rarely observed outside of winter since most of the Tibetan gazelle's range includes arid or semiarid areas with no soft ground surface. Tracks (<5 cm) are found after a snowfall and easily recognized as the smallest ungulate track on most of the plateau.

Old, dry fecal pellets

Tibetan Antelope

Pantholops hodgsonii // Zang Ling

Chiru

Adult male in winter

IUCN: NT
CITES: I
CPS: I
CRL: NT
BL: 100–140 cm

TL: 18–30 cm
WT: Male, 35–42
kg; Female, 24–
30 kg

Distribution

Tibetan antelope are mainly distributed in China across the Qinghai-Tibetan Plateau, extending to northwest India (Jammu-Kashmir). Within the region of this guide, Tibetan antelope are found in southern Qinghai Province and northern Tibet AR.

Appearance

The Tibetan antelope is a medium-sized antelope with a distinctive horn structure, enlarged snout, and dense, woolly coat. Males (35–42 kg) are much larger than females (24–30 kg). Their overall pelage is sandy brown to yellowish tan, and the ventral part is pale white. Adult males have a conspicuous black mask on their face with contrasting white eye-rings and upper lips. Males also have

Adult male in summer

Juvenile male in summer

black stripes on the front of each leg. The winter coat is smooth and much paler, appearing almost white, whereas the summer coat is coarse and tan. The tail is woolly and long, with a white rump patch underneath. Adult males have long, slender horns (50–70 cm in total length), growing vertically upward from the head with tips curving a little forward. The front of the horn is annulated with high ridges. When seen from the front, the two horns make a large V. Horns are absent in females.

Habitat and Ecology

Tibetan antelope are typically found in open habitat of grasslands, deserts, and steppes at high elevation from 3,700 m to 5,500 m across the Qinghai-Tibetan Plateau. Their diet includes grass, forbs, and lichens. Tibetan antelope are extremely wary animals. They are typically found in small herds of 10–20 individuals, but the adult males are usually solitary during nonbreeding seasons. They will form large herds of hundreds or thousands during migration. Females may travel 300–400 km annually from the wintering areas to their traditional birthing grounds and give birth to their offspring (typically a single calf) during May–June. Females and males segregate during migration, and the males may move only a short distance from the wintering areas. Some local populations remain sedentary and do not migrate. Rutting and mating occur in winter (November–December), during which the adult male will guard a harem of 10–20 females against other males. Fights between mature males are fierce and involve chasing and fighting and may result in fatal wounds. The calves are precocial and able to follow their mother about 1 hour after birth. Wolves are their primary predator. When approached or alerted, the calves will lie still on the ground. During the past several decades, Tibetan antelope have been heavily poached for their fine undercoat to make shahtoosh wool and for their horns for traditional medicine;

Track and scats

Fecal pellets

this poaching has caused severe population declines across their range.

Scat

The scats are typical goatlike black pellets. Separated fecal pellets are typically round to oval, pointed at one end and blunt at the other. The pellets are often found spread across a large area as a result of the individuals foraging as they defecate.

Tracks

Tibetan antelope primarily occur in arid open alpine habitats, so few tracks are observed outside of the winter season. Tracks can be found after a snowfall or occasionally on sandy ground.

Other Sign

While resting, Tibetan antelope may create a shallow pit (20–30 cm in depth) with their hooves. The animal will lie in the depression to rest and protect itself against harsh wind and snow.

Gaur

Bos gaurus // Yin Du Ye Niu

Adult

IUCN: VU
CITES: I
CPS: I
CRL: CR

BL: 250–330 cm
TL: 70–105 cm
WT: 650–1,500 kg

Distribution

Gaur are distributed from India, through part of the southern slope of the eastern Himalayas, into Southeast Asia. Within the range of this guide, gaur are found in southern Yunnan Province and probably southeast Tibet AR. Gaur may freely interbreed with other bovids, so hybrids are possible in the wild and under domestication. The semidomesticated hybrid of the gaur, called gayal or mithan, is locally common in southeast Tibet AR and northwest Yunnan Province (Dulong River and Nujiang River basins), extending to northern Myanmar and part of northeast India and Bangladesh. Free-ranging wild populations of gayals can also be found within this region and are referred to as *B. frontalis,* with the Chinese name "Da E Niu," and are evaluated as CR in the China Species Red List.

Appearance

Gaur are the second largest (after the Asian elephant) mammal within southwest China. They are large cattle with stubby legs. Adults are characterized by their conspicuous humped shoulders with massive muscles, especially in the males. The overall pelage is dark brown to black, with a dense and rather short coat, whereas the lower part of the four legs is contrasting white or gray yellowish, a white stocking pattern. Gaur have no white rump patch, a distinct character when compared with the banteng (*B. javanicus*), which historically occurred in southern China. Both sexes of gaur have one pair of horns, which grow from the side of the head and curve upward. The horns are yellow to tan at the base and turn black at the tips. The forehead between the two horns is gray to whitish with longer hairs, and the nose is also grayish to white. Gaur have a long tail, similar to an ox's tail, with brushy, long hairs at the end. The coloration of the gayal is more variable, depending on the crossbreeding. Gayals have a smaller body size and smaller shoulder bump compared with a wild gaur. Gayal horns are much less curved than those of the gaur and therefore form a broader and flattened forehead. Gayals also have a longer dewclaw than gaur, and the adult male usually has a larger dewlap on the throat.

Adult gayal (*B. gaurus frontalis*)

Habitat and Ecology

Gaur inhabit tropical forests in flat low-elevation areas (typically <1,000 m). They feed on grasses, young leaves of shrubs and trees, fruits, and bamboo. Gaur usually live in small herds of 3–12 individuals comprising adult females and their offspring of varied ages (juveniles and calves), whereas adult males are solitary and may sometimes form small bachelor groups. Larger herds of up to 40+ animals are occasionally observed. Gaur are mainly diurnal but may stay active throughout the day and night. The herds normally rest in shady places to avoid the midday sun. They will regularly visit water to drink and mineral deposits. Mating occurs during the dry season from November to March, during which the adult males will fight and compete for dominance of female herds. Females give birth to a single calf after a 9- to 10-month pregnancy. The gaur is an important prey species of tigers and leopards in term of biomass consumed. Human harvest is the primary current threat to the gaur throughout its range. Gaur have been heavily poached for meat, traditional medicine, trophies, and handicrafts (e.g., skulls and horns), both in China and elsewhere.

Tracks

Gaur have the largest hooved track besides domestic cattle and buffalo. Their tracks (10 × 13 cm) are larger and more rounded than those of other sympatric bovids (e.g., serows) and deer species. Their tracks are frequently found near waterholes and mineral deposits.

Scat

Gaur scat is similar to that of domestic cattle and buffalo. The most common form is a loose patty or a big single mass. Fresh scat varies from reddish brown to dark gray or black depending on diet, and old scat is normally black or dark brown.

Other Sign

Mating males make loud bellowing vocalizations that can be heard over a long distance. When alarmed or disturbed, gaur can make a resonant whistling snort. The vocalization of a cow-like "moo" is also commonly recorded. Their bedding sites are multiple shallow depressions on the soft ground, normally accompanied by their tracks and scat nearby.

Old, flat, pie-shaped scat

Hind foot track

Fresh, large mass scat

Takin

Budorcas taxicolor // Ling Niu

Adult female golden takin (*B. t. bedfori*)

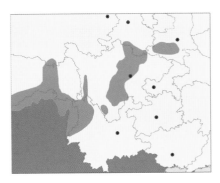

IUCN: VU
CITES: II
CPS: I
CRL: Listed as
   B. bedfordi,

B. tibetanus, and
B. whitei, all VU
BL: 190–210 cm
TL: 10–20 cm
WT: 200–600 kg

Distribution

Takins are distributed from central China, through the east flank of the Qinghai-Tibet Plateau, to the Himalayas (Bhutan, northeast India, and northern Myanmar). All four subspecies of takin found in China are within the range of this guide. Coat color differences exist between subspecies, but each subspecies produces the same track and sign across the range. The distribution of each subspecies in China is as follows: the Qinling Mountains in southern Shaanxi Province for *B. t. bedfordi* (golden takin); the east edge of the Qinghai-Tibet Plateau, including southern Gansu and central Sichuan Provinces, for *B. t. tibetana* (Sichuan takin); part of southeast Tibet AR and northwest Yunnan Province for *B. t. taxicolor* (Mishmi takin); and Tibet AR for *B. t.*

whitei (Bhutan takin). The geographic divide between *B. t. taxicolor* and *B. t. whitei* is not well determined.

Appearance

Takins are large robust ungulates with higher stature at the shoulder than the hip. They have a large, stocky head, and the face in profile is distinctively convex. Adult females are smaller than adult males. Both sexes have black to brown horns, which arise vertically as yearlings, then turn abruptly backward and slightly upward as the animal matures. The horns of adult males are more separated and thicker near the base than those of females. Takin have a dense, wooly coat with long and coarse hairs, normally with a darker stripe in the middle of the back. Coat color for adult takins is yellow or brown in general but is highly variable with distinct variation among subspecies, from the completely golden yellow of *B. t. bedfordi* in the north to the yellow brown with black patches of *B. t. tibetana* in the central region to the black or dark brown of *B. t. whitei* in the south.

However, even within populations there is great variability in coat color. The pelage of females and yearlings is usually grayer than that of males. Adult males may have a conspicuous mane with long, dark reddish or brownish shaggy hairs, especially during the rutting season (summer). Calves of all subspecies are dark brown throughout, with a distinct stripe on the back. Takins have broad hooves and well-developed dew toes.

Adult male golden takin (*B. t. bedfordi*)

Habitat and Ecology

Takins inhabit temperate montane forest, shrubland, and alpine meadows. They are agile climbers; although they usually move slowly, they are capable of quick movements when charging rivals or perceived threats. Takins occupy a broad elevation range between 1,000 and 4,200 m. Some takin populations conduct seasonal vertical movements. For Sichuan takins, most individuals are found in alpine meadows above the tree line during summer and move down to lower forest and valleys in fall when the forage at high elevation begins to decline following the first frost. They spend most of the winter in forests with bamboo understory at the medium elevation and move to the lowest-elevation valleys in early spring, seeking newly emerged vegetation. We are not aware of seasonal movement patterns for animals living outside of the range of bamboo forests, such as those takins in rhododendron forest. The takin diet includes grasses, forbs, bamboo shoots and young woody twigs, new leaves of shrubs and trees, and tree bark. Takins can cause substantial damage to shrubs and tree saplings while foraging since they will bend or break the branches with their body

Adult Bhutan takin (*B. t. whitei*)

Adult female Sichuan takins (*B. t. tibetana*) with calf

to reach young leaves. Takins will regularly visit mineral licks. They can move in herds of 10–30 individuals, consisting of multiple adult females and their offspring, as well as juveniles of both genders. Large herds of up to 300 animals are occasionally found in open meadows above the tree line during summer but are not stable. Adult males will temporally join the group during the mating season (June to late July or early August) but usually remain solitary. Males exhibit intense fighting during the rut, which consists of short-distance charges with a lowered head and pushing, and these contests are sometimes fatal. Females give birth to young, mostly singles but occasionally twins, in spring (March–April). When the herd is moving, all calves may follow a single adult female. Despite the season, takins maintain a similar daily activity pattern of foraging during early morning and late afternoon and resting and ruminating the rest of the day. Solitary adult males are aggressive and will charge people when they perceive a threat, as will adult females with young. Few natural predators remain in their range, but bears and small carnivores will feed on carcasses, which are obvious in stream beds each spring.

Scat

The scat of takins varies broadly by season depending on diet. During winter and early spring, takins produce oval or round brown-black pellets (2–3 cm), although the color may vary from reddish brown to dark gray or black. Sometimes the pellets are in large clusters. In summer and fall when they feed more on forbs, scats can be a loose patty or a single mass.

Tracks

Within their habitat, takins leave the largest hooved track besides that of domestic cows and yak. Their tracks (9 × 10–12 cm) are more rounded in the front than those of other sympatric bovids and deer species. The print of the well-developed dew toes is frequently visible when tracks are left on soft or sandy ground.

Other Sign

Takins peel soft tree bark using their lower incisors. Such activity is seen primarily in the spring, possibly to access tree sap or feed on the succulent cambium. These scrapes can occur up to 2 m aboveground, caused by the takin standing on its hind legs. Takins frequently rub their horns and body against trees or stumps, leaving rub marks. If there are large deer species (e.g., sambar deer) co-occurring with takins, the rub marks may be indistinguishable; however, long, coarse hairs of takins can frequently be found at these rub sites, which could be used for identification. Mineral licks are regularly visited by takins, and large amounts of scat, tracks, and feeding signs can be found there. One common type of mineral lick is smooth, compressed soil in a depression on a bank or slope. Teeth marks are common on the soil. Mineral licks are often visited by multiple ungulate species.

Fresh separated fecal pellets

Dry clustered scat

Fresh loose-patty scat

Track

Hairs

Rub marks

Feeding sign

Teeth marks

Kiang

Equus kiang // Zang Ye Lv

Tibetan Wild Ass, Wild Ass

Mother-foal group

IUCN: LC
CITES: II
CPS: I
CRL: NT

BL: 180-220 cm
TL: 32-45 cm
WT: 250-400 kg

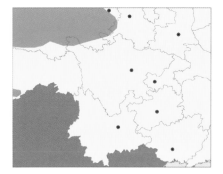

Distribution

The kiang is widely distributed across the Qinghai-Tibet Plateau, except for the southeast region, and the western Himalayas. Within the region of this guide, kiangs are found on high plateau in southern Qinghai and southwest Gansu Provinces.

Appearance

The kiang is a robustly built muscular equid with a large head and blunt muzzle and also the only equid species within the region except for domestic horses. The dorsal side is brown to chestnut; the ventral side and legs are whitish to gray, with a distinct edge across both flanks. The kiang's summer coat is short and sleek, whereas the winter coat is thick and shaggy, and the color is usually darker. Kiangs have an erect black mane followed by a dark dorsal stripe extending to the tail. Their ears are black tipped.

Habitat and Ecology

Kiangs prefer open habitat with abundant grasses and sedges on the plateau, including alpine meadows, plains, and broad valleys. They are also found in arid habitats such as desert steppe and dry basins. Kiangs occur within a broad elevation range from 2,700 to 5,400 m. Their diet is coarse, fibrous vegetation like many other wild equids. Kiangs are vigilant to any moving objects on the open terrain but are highly curious. They may run after vehicles and slowly approach humans to inspect them. Kiangs do not maintain permanent social groups, although large herds containing hundreds of individuals are occasionally observed, especially during fall and winter. Young males may form a bachelor group of multiple individuals. Adult males will form harems during the summer mating season and will guard their females against intruders. Foals are mostly born in summer (July–early September). Kiangs may conduct seasonal movements to pursue favorable forage, but there are no regular migration patterns.

Scat

Kiang scats are similar to those of domestic horses and distinct from those of all other even-toed ungulates on the plateau. Their scats are normally large, round or oval, flat pellets (5–7 × 3–5 cm) or oval pies containing mostly undigested grass. Fresh scats are yellowish to brown with a smooth, moist surface, and old scats are dry and dark brown to black. Kiang scats are often found in large piles where the herd has foraged.

Tracks

Kiangs leave a typical equid track with an oval print that is open in the back (8.5 × 6 cm), with the front hoof print being slightly larger than the hind hoof print. Clear tracks are difficult to find on grass meadows or arid habitats but are observable on snow or moist ground.

Scat

Dry fecal pellets

Large foraging herd

Wild Boar

Sus scrofa // Ye Zhu
Wild Pig, Eurasian Wild Pig

Adult male

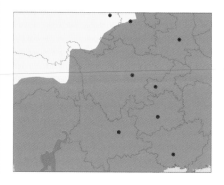

IUCN: LC
CITES: NL
CPS: NL
CRL: LC

BL: 100–150 cm
TL: 17–30 cm
WT: 50–200 kg

Distribution

Wild boars have one of the widest global distributions among terrestrial mammals. They are found throughout southwest China except in the high plateau (normally <3,500 m elevation).

Appearance

The wild boar is a large, robust suid, whose appearance is similar to that of a domestic pig but with an elongated muzzle, thicker hair, and darker coat. Its body color varies from dark gray to brownish and black. Adults have long hairs on the back and neck, and adult males have large, strong lower canines (tusks) protruding on the sides. Wild boar piglets are brown with white or pale yellow stripes along the body that gradually disappear during the first year.

Habitat and Ecology

Wild boars are adapted to various habitats, including forest, shrubland, tree plantations, grassland, and the forest-agriculture matrix. They are omnivorous and feed on tubers, grass, fruits, mast seeds, crops, invertebrates, and small vertebrates. They will also scavenge on animal carcasses. They are generally gregarious, with loose social organization, as they are observed as either solitary individuals, mother–young units, or herds. They have a high reproductive rate, with a litter size 5–10+, and females breed up to twice per year. Wild boars are important prey for large predators and are important seed dispersers. They can breed with domestic pigs and are a frequent source of wildlife-human conflicts because they can cause extensive damage to crops or tree plantations.

Scat

Wild boar scats vary greatly in form and shape depending on diet, from dry pellets (rare) to soft, segmented cylinders with blunt ends (common; diameter varying from 3 to 5 cm). Formless scats of a loose single mass are also commonly found. The scat color varies from dark gray to black. Fresh scats usually have a strong odor. Regardless of the form and shape, wild boar scats normally contain obvious plant fibers because of their rooting behavior. Mast shells, berry seeds, grass, and animal hairs and bones can also be identified. Sometimes wild boar scats are not easily distinguished from those of Asiatic black bears, but bear scats usually contain less fibrous material.

Tracks

Wild boars leave typical tracks of even-toed ungulates, but their tracks (6 × 5 cm for the forefoot, 6 × 4 cm for the hind foot) are more rounded than those of cervids and musk deer. When found on soft ground (wet soil or mud), tracks have imprints of the two dewclaws. The two hooves are more widely separated on the forefoot than the hind foot.

Fresh scat

Dry fecal pellets

Adult visiting a mud wallow

Other Sign

Wild boars frequently dig for plant roots and other food with their snout, which results in a large number of shallow digging pits with obvious soil disturbance. Bedding sites can be found in forests and are commonly built for one-time use and composed of bamboo or soft shrub branches cut by their teeth. Wild boars will use these materials to cover their body while resting. The bed is normally round or oval and looks like a huge ground nest up to 2 m in diameter. Wild boars frequently create mud wallows along the edges of small ponds or streams. Large numbers of tracks and body prints can be found in and around the wallow.

Tracks

Digging pit in open meadow

Digging pit at tree root

Bedding site

PANGOLIN // CHUAN SHAN JIA

Chinese Pangolin
Manis pentadactyla // Chuan Shan Jia

Adult

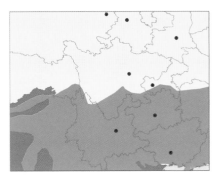

IUCN: CR
CITES: II
CPS: II
CRL: CR

BL: 42–92 cm
TL: 28–35 cm
WT: 2.4–7 kg

Distribution

This species occurs in South and Southeast Asia. In China the species occurs south of the Yangtze River. In our study area, the distribution includes Yunnan and Guizhou Provinces and Guangxi AR, as well southern parts of Sichuan Province, Chongqing City, and Tibet AR. Currently, it is extremely rare and may be extirpated from much of this range.

Appearance

The Chinese pangolin has the appearance of a scaly anteater. The streamlined body and long, flat tail are covered with rows of overlapping brownish scales. The soft, off-white ventral side and much of the face, including the visible ears, are not covered. When a Chi-nese pangolin feels threatened, it curls its entire body into a ball. The species has long, thick front claws.

Habitat and Ecology

This species is found in a wide range of habitats, including primary and secondary tropical forests, limestone forests, bamboo forests, broadleaf and coniferous forests, grasslands, and agricultural fields. The Chinese pangolin is solitary and primarily nocturnal (sometimes crepuscular). It is largely terrestrial, although it is fully capable of climbing trees and, like other pangolins, swims well.

Scat

Pangolin scats are composed almost entirely of ant or termite exoskeletons. No other mammal of that size in the study area eats ants exclusively. The scats are compact and spherical or cylindrical, with the diameter varying from 2 to 4 cm. Scat color varies from brown

Burrow

Burrow

to black. Fresh scats are rarely seen because pangolins are reported to bury their feces after excretion. Regardless of the form and shape, pangolin scats normally contain obvious insect exoskeletons and dirt because of their feeding behavior.

Tracks

Pangolin tracks have five toes on each foot. The first and last digits on the forefeet are reduced, whereas the middle three digits are well developed. Pangolins leave round-shaped prints for the hind foot (6 to 8 cm) and three claw marks for each front foot, with signs of the tail being dragged.

Other Sign

Because of their habit of nocturnal activity and diurnal resting within a burrow, sightings are rare. Extensive digging around ant and termite mounds is usually attributed to pangolins but can be Asiatic black bears. Pangolins use feces, urine, and scent gland secretions to mark their territory, but these sign are not obvious.

Asiatic Brush-Tailed Porcupine
Atherurus macrourus // Zhou Wei Hao Zhu

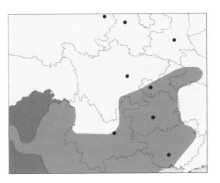

Adult, captured by camera with infrared flash

IUCN: LC
CITES: NL
CPS: NL
CRL: LC

BL: 34–53 cm
TL: 14–23 cm
WT: 2–4 kg

Distribution

The Asiatic brush-tailed porcupine is distributed in subtropical and tropical montane forests of northeast South Asia and Southeast Asia and into central China. In the range of this guide, it is distributed in Guizhou Province, Chongqing City, and Guangxi AR, as well as eastern Sichuan and Yunnan Provinces.

Appearance

The Asiatic brush-tailed porcupine is a small-sized, slender porcupine. Quills cover almost the entirety of its body, although quills are soft on the underparts, head, and legs. Fur on the upper back is blackish brown to grayish brown, and it is white to light brown on the ventral side. The species possess-

es short and stout limbs, with short, rounded ears. The tail is pale and covered with small scales, and the end is a brush-like arrangement of hairs.

Habitat

Asiatic brush-tailed porcupines are terrestrial and nocturnal, inhabit subtropical and tropical montane forests, and avoid deforested areas up to 3,000 m elevation. The species is often found in areas with diverse understory plantations such as cane and bamboo. These animals may dig burrows in soft soil and are often found near water.

Scat

Scats are similar in shape to those of the Malayan porcupine but are smaller in size (0.5 cm diameter).

Tracks

Their forefoot (4 × 3 cm) and hind foot (7 × 4 cm) tracks are five toed and show blunt, straight claws. However, the thumb toe imprint is often missing on the forefeet, leaving only four claw marks. Toes are widely spread when walking. The soles are naked and are fitted with pads. The feet turn in as the porcupine walks.

Other Sign

Each quill is thick and scaly and contains a chain of flattened disks that rattles when shaken. This sound may serve to deter predators.

Malayan Porcupine

Hystrix brachyura // Hao Zhu

Adult

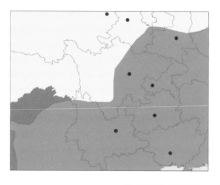

IUCN: LC
CITES: NL
CPS: NL
CRL: LC

BL: 56–74 cm
TL: 8–12 cm
WT: 9–14 kg

Distribution

This species ranges throughout central and southern China. In southwest China, it is distributed in all provinces east of the plateau region (therefore excluding Qinghai Province, western Sichuan and Gansu Provinces, and the Tibet AR). Recent literature speculates the porcupines within the range are two separate species, Malayan porcupine *H. brachyura* and Chinese porcupine *H. hodgsoni*.

Appearance

The Malayan porcupine is one of the largest of Southeast Asia's seven species of porcupines. The front half of the body is dark brown to near black. The rear half is equipped with long, sharp quills, which are banded black and white or dark brown and white. Typically, the longer quills are predominantly white, with a dark band in the middle. Long, thick hairs on the nape may be erected into a crest.

Habitat and Ecology

Malayan porcupines are found in various forest types, open areas near forests, and agricultural areas up to

3,000 m elevation. The species is mainly terrestrial and sleeps or rests in tree cavities or burrows. The burrow is usually self-constructed, with an entrance tunnel (entrance 30 × 30 cm), multiple exits, and a large inner chamber.

Scat

The scat is shaped like a cashew nut (length of 2–3 cm, width of 1 cm); many scats are curved with a subtle point at one end, and the ends are smaller than the middle of the scat. The dark green pellets are clumped when fresh but form separate brown pellets when dry. Scats can be composed of dense fibrous plant material, especially in winter when Malayan porcupines feed on the dry material of woody plants. Sometimes porcupine scats can be found in large piles.

Tracks

The species has an oval-shaped print with four toes on the fore print (5 × 3 cm) and five toes on the hind print (8 × 4 cm). The claw may be obvious in the print. The track of the hind foot is larger and slenderer than that of the forefoot. The footprints turn inward along the path of travel.

Scat

Habitat

Other Sign

The Malaysian porcupine digs burrows at the base of trees and inhabits dens near rocky areas. When using a burrow, a network of trails usually radiates into the surrounding habitat. It molts its quills in the spring, and they can be found throughout its home range. When feeding on twigs, the animal leaves twigs with multiple small bite marks. Sometimes it feeds by peeling the bark from tree trunks.

Scat pile

Feeding site

Quill

Black Giant Squirrel

Ratufa bicolor // Ju Song Shu

Adult

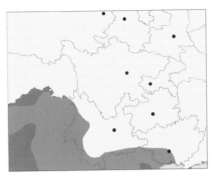

IUCN: NT
CITES: II
CPS: II
CRL: VU

BL: 36–43 cm
TL: 40–51 cm
WT: 1.3–2.3 kg

Distribution

This widespread species ranges from northern South Asia, through southern China, into much of Southeast Asia. In China, the species has been recorded in southern Yunnan Province, eastern Tibet AR, southern Guangxi AR, and Hainan Island.

Appearance

The species' dorsal pelage and tail are deep brown to black, but the cheeks, chest, front of the forelimbs, and ventral surface are cream or orange. The tail is long, black, and dorsal-ventrally flattened.

Habitat and Ecology

The black giant squirrel is a diurnal and arboreal species occasionally feeding on the forest floor. It occurs in tropical and subtropical montane evergreen and dry deciduous forests up to 1,400 m elevation, nesting in tree hollows in the mid-canopy or higher.

Scat

Black giant squirrel scats may be found on the ground under their feeding sites. Their tubular scat is variable in shape,

Feeding site

color, and length, but the width is about 2 cm, and it has an indistinct dimple at one end and no point at the other. The size of their scats is larger than those from other squirrels.

Tracks

The tracks of black giant squirrels are rarely seen because of their arboreal habits. These squirrels generally remain high in the canopy but can be observed feeding at lower levels. The tracks of the forefeet are about 2 × 2 cm, and the hind feet are longer, about 7 cm long by 2 cm wide. The hind prints show five toes, whereas fore prints show four to five toes. The claw marks are visible, and footpads are usually visible when tracks are created in mud.

Other Sign

Partially eaten nuts, fruits, and leaves can be found on the ground below feeding sites.

Himalayan Marmot
Marmota himalayana // Xi Ma La Ya Han Ta

 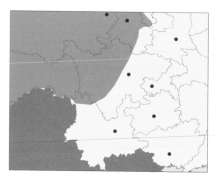

Pair in alpine grassland habitat

IUCN: LC
CITES: III
CPS: NL
CRL: LC

BL: 48–67 cm
TL: 12–15 cm
WT: 4–10 kg

Distribution

This species is present in western, southwest, and central China. In our study region it is found in the plateau regions of Qinghai, western Gansu, Sichuan, and Yunnan Provinces and Tibet AR.

Appearance

The Himalayan marmot is about the size of a large housecat. It has a dark chocolate-brown coat with contrasting yellow patches on its face and chest. Individuals have distinct dark brown patterns on their lips, nose, forehead, and spine and the tip of tail. Male marmots are larger and heavier than females.

Habitat and Ecology

This species is found in alpine meadows, grassland, and deserts with very low annual rainfall, typically inhabiting either steep rocky slopes or gentle slopes where soil can be readily excavated. It lives in colonies, with colony size depending on local food resources. Himalayan marmots excavate deep burrows, which are shared by colony members during hibernation. Sometimes these marmots live around human houses and establish a close relationship with humans.

Scat

Latrines are often found within Himalayan marmots' home range. The size and shape of the scat will vary with the diet and size of the marmot, and the pellets are either separate or clumped. Separate pellets are tubular shaped (5–7 × 1–2 cm), pointed on one end and round on the other. Scats are composed of fibrous plant materials, seeds, and shells.

Tracks

Fresh tracks can be found on mud, sand, or snow. The hind print is about 8 cm long with five toe pads; the fore print (5 cm long) shows only four toe pads. Claw marks on the hind print are more widely spread than those on the fore print.

Track

Habitat

Scat

Other Sign

Himalayan marmots live in burrows with multiple wide and smooth entrances (40 × 30 cm, sometimes within rock piles). They excavate deep burrows that colony members share during hibernation and may excavate temporary shallow burrows in summer. They are highly social and use loud whistles to communicate. One marmot will stand at the burrow entrance and use the loud "ku-bi" whistle to alert colony members when alarmed.

Burrow entrance

Adult in typical alpine rocky habitat

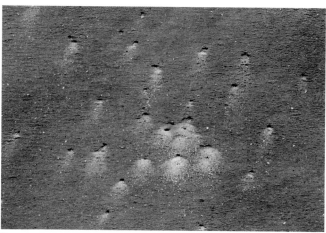
Burrows

HARES // YE TU

General Description
Hares (*Lepus*) are the larger species found within the family Leporidae. They have long ears, a divided upper lip, and long hind limbs adapted for leaping. Hares are herbivores and fast runners and typically live solitarily or in pairs. Newborn hares (leverets) are born fully furred with eyes open.

Habitat and Ecology
Hares live primarily in open fields with scattered brush or early successional trees for shelter. They are very adaptable and thrive in mixed farmland, semidesert, alpine meadow, and residential areas. In contrast to rabbits, hares do not burrow underground. They usually bear their young in a shallow depression or flattened nest of grass called a form.

Yunnan hare habitat

Scat

Hares produce feces that are consistent in shape and color. Scat pellets are round in shape and small in size (0.5–1.5 cm diameter). Scats are composed of compacted greenish (fresh) to brownish (old) pellets of fibrous plant material.

Scat of Yunnan hare

Tracks

The hind foot track is generally rectangular to triangular in shape, whereas the print of the forefoot is more circular or oval shaped. For tracks in dense snow or mud, four toes can be distinguished on both the fore and hind feet. The tracks of the hind feet are larger (length of 10–15 cm, width of 5–8 cm) than those of the front feet (length of 5–8 cm, width of 3–5 cm). Hares generally travel in a bounding gait as they move, and usually, the hind feet register ahead of the forefeet. The distance between sets of tracks in a trail might be a few centimeters to over several meters, depending on the speed of movement.

Other Sign

It is common to find feeding sign when hares are present; usually, these are cuttings on leaves (clear-cut, not jagged), twigs (clear-cut), and small tree sapling trunks (distinct teeth marks).

Chinese Hare

Lepus sinensis // Hua Nan Tu

In winter coat

IUCN: LC
CITES: NL
CPS: NL
CRL: LC

BL: 35–45 cm
TL: 4–6 cm
WT: 1–2 kg

Distribution

This species occurs in southeast China to the south of the Yangtze River. In our study area it is distributed in Guangdong, Guizhou, and Yunnan Provinces and Guangxi AR.

Appearance

The Chinese hare is a small species, with the adult females being larger than the males. The fur is short and coarse; the dorsal side and chest pelage is brownish, and the ventral side is light gray. The tail is brown, and the tips of the ears bear triangular black patches.

Tolai Hare

Lepus tolai // Meng Gu Tu

Adult

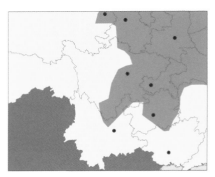

IUCN: LC
CITES: NL
CPS: NL
CRL: LC

BL: 40–59 cm
TL: 7–11 cm
WT: 1.6–2.7 kg

Distribution

The Tolai hare occurs in western, central, and northeast China. In our study area, it is found in eastern Qinghai Province, Gansu and Shaanxi Provinces, eastern Sichuan and northern Yunnan Provinces, and northern Guizhou Province.

Appearance

Tolai hares are large bodied. The dorsal pelage from the nose to the back varies from yellow white to yellow brown. The short tail is wide and pale colored, with narrow black stripes. The abdominal fur is mostly pale or white.

Woolly Hare

Lepus oiostolus // Gao Yuan Tu

Adult in winter coat

IUCN: LC	BL: 40–58 cm
CITES: NL	TL: 6–12 cm
CPS: NL	WT: 2–4.5 kg
CRL: LC	

Distribution

The woolly hare is a high-altitude lagomorph native to the Tibetan plateau. The distribution of the woolly hare includes western and central China, including the plateau regions of Gansu, Qinghai, Sichuan, and Yunnan Provinces and Tibet AR.

Appearance

Woolly hares are relatively large compared with other hares. The dorsal pelage from the nose to the back is slightly wavy, and the color varies from yellowish white to brown. The rump varies from brownish gray to silver gray, and the short tail can be white or white with gray or black stripes. Abdominal fur is mostly white and may have a light brown line along the mid-ventral line. Adult female woolly hares are bigger than males.

Fleeing adult in winter coat

Yunnan Hare

Lepus comus // Yun Nan Tu

Adult

IUCN: LC
CITES: NL
CPS: NL
CRL: NT

BL: 32–48 cm
TL: 9–11 cm
WT: 1.8–2.5 kg

Distribution

The geographic distribution of the Yunnan hare includes Yunnan, western Guizhou, and southern Sichuan Provinces in China. It is found on the Yunnan-Guizhou Plateau at elevations of 1,300–3,200 m.

Appearance

Yunnan hares are medium sized. The dorsal pelage is relatively long and soft, and the color varies from light gray to dark gray. The color of the tail is light brown, and can be light gray on the ventral side. A pale strip extends from the nose to the ear, with a bow-shaped curve on the brow.

BIRDS OF SOUTHWEST CHINA

General Description

China harbors a high diversity of species within the order Galliformes (i.e., pheasants, grouse, partridges, peafowls, and their related species), which are referred to here as galliforms. The galliform species in China (63) account for 22% of the world's species. Phasianidae (e.g., pheasants, quails, and partridges) are especially diverse, with 55 of the 159 species found in China.

Southwest China maintains a high species richness and endemism of galliforms. For the 47 galliform species found within the geographic range of this guide, 14 are endemic to China. These birds inhabit diverse habitats, from tropical forest in the south to alpine meadows and mountains within the Tibetan plateau. Their body weights vary from 70 g to 5 kg; most of them are large and attractive and have long been linked to humans. Currently, these large birds are undergoing severe threats due to habitat loss and fragmentation, as well as illegal harvest. Seventeen (36%) of these species are considered globally threatened (NT, VU, EN) by the IUCN, and more than half (55%) of them are on the list of China's national protected animals.

We have included galliforms in this wildlife guide because they are one of the unique types of wildlife of this region, and they are relatively easier to observe than many large mammals. They can be very tame in the Tibetan regions where, traditionally, people do not hunt, and some birds are fed by monks and visitors around the monasteries. The birds also

Monks feeding birds

leave visible signs, such as nests, eggs, droppings, feathers, etc. But unlike the large mammals, most galliform signs cannot be identified with specific species. We describe the most encountered signs and the nests and eggs for each species when information and photographs are available.

Galliformes is a unique order of birds. Their wings are short and round, and their legs are strong, with sharp claws, making them better at running than flying. They have short, robust bills that are good at pecking and digging. They feed mainly on plant materials and occasionally on invertebrates (e.g., insects and worms), and a few feed on small vertebrates; the chicks consume more animal material than the adults. Their social organization ranges from monogamous to polygamous. Species with marked sexual dimorphism tend to be polygamous; their males have developed complicated courtship displays, and only the males have spurs in most species. All species nest on the ground, except for the tragopans, which primarily nest in trees. Their nests are simple, normally a shallow pit on the ground filled with leaf litter and downy feathers. The precocial chicks are capable of moving around and feeding themselves shortly after hatching.

In this guide we describe 31 species that weigh more than 500 g and occur regularly within the geographic range of this guide. Galliformes is a specious order in this region; we could not cover all species in the limited length of this guide, so we concentrated on the most common and visible species. Several species, the Himalayan snowcock (*Tetraogallus himalayensis*), chukar (*Alectoris chukar*), Daurian partridge (*Perdix dauurica*), Cabot's tragopan (*Tragopan caboti*), and Elliot's pheasant (*Syrmaticus ellioti*), are excluded because they occur primarily outside of our focal range. All species of the genera *Coturnix* and *Arborophila* are excluded because of their small body size and/or low visibility. For a complete list of species found in the region we recommend the reader consult taxonomic volumes such as MacKinnon and Phillipps (2000). The range of several species is not well defined, as there are variations between IUCN records and recent guides. In most cases we present the IUCN range map, but we indicate when we have modified the map to fit our knowledge of the region.

Droppings

Droppings (i.e., feces) are the most common sign of wild galliforms. They have a curled shape similar to that of domestic chickens, but the size varies between species. They are in most cases scattered but can be in small piles when under a roosting site. Fresh droppings are usually dark green to gray, with one-third to one-half of the length covered by a thick white layer. Old droppings turn yellow to brown in color. The white layer is composed of metabolites, commonly uric acid. The white layer may be absent when the dropping contains a high water content, in which case it is a formless jellylike mass or thick liquid that gradually turns dark brown to black. Birds feeding on young plant materials often produce these alternative droppings.

Fresh formless dropping of buff-throated partridge. Note the usual white uric acid layer is absent.

Fresh droppings of golden pheasant

Fresh droppings of Temminck's tragopan

Dried dropping of white eared pheasant, exposed to direct sunshine

Tracks

The tracks of galliforms are rarely found unless they are fresh and left on a soft surface, such as in the sand, mud, or snow. A print includes three toes pointing forward, and the one pointing backward is usually reduced or absent. Their tracks normally present as a walking pace rather than the typical hopping pace left on the ground by other non-galliform birds such as magpies, thrushes, and laughing thrushes.

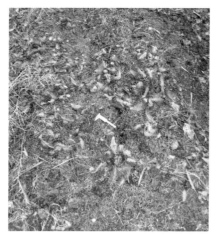

Feathers of male Temminck's tragopan   Digging pit left by golden pheasant

Feathers

Molted feathers of galliforms are frequently found in their habitat, typically at their nest, roosting, and sand bath places. The tail, flight, and contour feathers are usually easy to identify with specific species on the basis of the feather's color and pattern, especially for the males of sexually dimorphic species. Feathers and body remains left by predators (e.g., carnivores and raptors) can be occasionally found on the ground.

Feeding Sign

Feeding sign are commonly found as the birds dig and scratch for food. The most common feeding sign are the digging pits, as indicated by disturbed leaf litter and soil. But such digging pits usually cannot be distinguished from the feeding sign left by sympatric mammals (e.g., hog badgers).

Sand Bath Sites

Like many other bird species, galliforms frequently take a sand bath to clean their body and feathers of ectoparasites. Each flock may have their conventional bath site, which is typically a small open ground, either covered or not covered by vegetation, with loose dirt or sand. Freshly used bath sites can be recognized as one or more shallow pits or depressions on the ground with disturbed dry dirt and sand. Such bath pits are distinctive from their feeding and digging pits by the larger size and conspicuous large amount of disturbed dry soil. The baths are normally taken around noon through afternoon in sunny days, and molted feathers can often be found at or near these sites.

Female Chinese monal taking a sand bath

Sand bath site of Hume's pheasant

Blood Pheasant
Ithaginis cruentus // Xue Zhi

Adult male from Sichuan Province

IUCN: LC	BL: 40–48 cm
CITES: II	TL: 14–18 cm
CPS: II	WT: 400–600 g
CRL: NT	

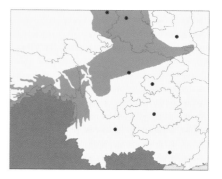

Distribution

The blood pheasant is distributed along the eastern Himalayas and in the montane areas on the Qinghai-Tibetan Plateau and in central China. Within the range of this guide, it resides in southern Gansu and Shaanxi Provinces, eastern Qinghai Province, western Sichuan and Yunnan Provinces, and the southern Tibet AR. The range shown is increased from that indicated by IUCN on the basis of our local knowledge.

Appearance

The blood pheasant is a medium-sized pheasant with sexual dimorphism. The male has a short black bill, red orbital skin, and a small crest on the head. The feathers are lanceolate shaped, and

plumage is streaked with white throughout, with gray on the dorsal and green along the ventral side. The feet and the vent feathers are bright red. Subspecies of the males are highly variable in coloration; the major difference lies in the amount of red and black on the head, throat, and breast and the color of the

Adult male from northwest Yunnan Province

Adult males from southeast Tibet AR

wing coverts (rufous, green, or gray). The female is gray or dull brown, crested, with red orbital skin and red feet.

Habitat and Ecology

Blood pheasants inhabit mountainous areas primarily in conifer and mixed conifer-broadleaf forests and alpine shrubs from 2,100 to 4,600 m elevation. They feed on leaves, buds, flowers, seeds, roots, mosses, mushrooms, and insects. The species is monogamous and breeds from April to July. Individuals that do not form pairs form small flocks during the breeding season. Foraging groups during winter can have up to 80 individuals.

Nest and Eggs

The nests of blood pheasants are found on the ground at the base of trees, under shrubs or grass, or in rock cavities. The nest is about 20–26 cm in diameter and 5–9 cm in depth. The clutch size is 4–10, and eggs are pale brown with dark brown speckles. The female incubates the eggs alone, and the male helps to fledge young. Incubation lasts about 24–37 days.

Other Sign

The droppings of blood pheasants are normally curved segments with a uniform diameter of 0.4–0.6 cm with blunt ends, distinct from that of other sympatric pheasants and grouse. The droppings usually contain coarse plant fibers and debris without a smooth surface. Fresh droppings are green, whereas the old, dry ones are tan to yellow or brown.

Adult male from Qionglai Mountains, central Sichuan Province

Adult female and juvenile

Nest

Fresh droppings

Old dry droppings

Blue Eared Pheasant

Crossoptilon auritum Lan Ma Ji

Adult

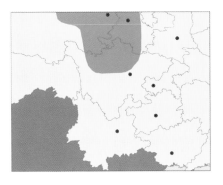

IUCN: LC
CITES: NL
CPS: II
CRL: NT

BL: 75–103 cm
TL: 40–57 cm
WT: 1.5–2.1 kg

Distribution

Blue eared pheasants are endemic to central and southwest China. Within the range of our guide, blue eared pheasants reside from northern Sichuan Province through eastern Qinghai and western Gansu Provinces

Appearance

Blue eared pheasants are a large, gray-bluish pheasant. The plumage of the species is monomorphic, although the male is larger than the female and has short spurs. They have a black cap, extensive red bare facial skin, and a white moustache stripe extending into the long ear tuft on each side. There is a white patch behind the hip, and the legs are red to bright pink. They have long, loose tails that are curved and dark tipped. The outer tail feathers are metallic purple blue.

Foraging flock

Habitat and Ecology

Blue eared pheasants inhabit conifer forest, alpine shrubs, and meadows across an elevation range of 2,400–4,000 m. They feed on plant materials, including leaves, buds, flowers, roots, and seeds, and some insects. The species is monogamous and breeds from late March to July, during which large flocks split into family pairs. Loud calls can be heard from a long distance during the breeding season, and the males engage in dramatic courtship displays and duels for females; their fights may continue despite severe bleeding wounds. Foraging groups of up to 30 individuals are observed outside of the breeding season. Large flocks of 100+ birds are occasionally seen during winter and early spring. Once disturbed, they escape by running rather than flying.

Nest and Eggs

The nests are constructed on the ground under shrubs or rock overhangs, occasionally in tree hollows near the ground. The nest is about 27–30 cm in diameter and 4–8 cm in depth. The clutch size ranges from 5 to 12, and eggs (about 5.6 × 4 cm) are incarnadine with tiny light brown speckles. Incubation takes 25–26 days and is undertaken solely by females.

Nest

Common Pheasant

Phasianus colchicus   Huan Jing Zhi

Ring-Necked Pheasant

Pair of the eastern subspecies

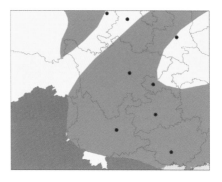

IUCN: LC	BL: 50–90 cm
CITES: NL	TL: 29–59 cm
CPS: NL	WT: 0.5–3 kg
CRL: LC	

Distribution

The common pheasant is the most abundant and widespread pheasant of the world; it is widely distributed in Asia and has been introduced to other continents as a game bird. Nineteen out of its 30 subspecies occur in China, distributed throughout the country except for Hainan Island and the high-altitude regions on the Qinghai-Tibet Plateau. Within the range of this guide, they reside throughout the region.

Appearance

The common pheasant is a large, bulky pheasant with sexual dimorphism. The male has red facial skin, and some have red wattles. The male has a metallic dark blue neck, copper-and-gold body plumage, and a long brown tail barred with black. Males of different subspecies vary in colors of body plumage. Notable differences include the presence or absence of the white neck ring and the white eyebrow, the width of the neck ring, and the color of the wing coverts (white to yellowish brown) and the rump (bluish gray to reddish). The bird is also called a ring-necked pheasant because a number of subspecies have a white neck ring. The female is brown, with light brown barring all over its body.

Habitat and Ecology

The common pheasant is a generalist and inhabits farmlands, shrubs, grasslands, wetlands, and forests, but it is most common in open fields with a small copse of trees. In Sichuan Province it is found up to 3,000 m elevation. Plant materials constitute the bulk of its diet, which also includes insects, invertebrates, and small vertebrates. The species is polygamous and breeds

Adult male of the western subspecies

Adult female

from March to July. The male establishes its breeding territory, and then the female selects the male to breed and stays within its territory. Winter flocks of the species often are segregated by sex. When startled, they often burst into flight from bushy cover, making heavy wingbeats and harsh "kok kok kok" calls.

Nest and Eggs

The common pheasant nests on the ground under dense vegetation cover and also in agricultural fields. The clutch size ranges from 4 to 8 eggs in southern areas and increases up to 20 eggs in northern areas in China. The eggs can be pale yellow, light gray, light brown, or pale olive, with no speckles. The female incubates the eggs alone for 22–26 days.

Golden Pheasant

Chrysolophus pictus Hong Fu Jin Ji

Adult male

IUCN: LC
CITES: NL
CPS: II
CRL: NT

BL: 60–110 cm
TL: 38–75 cm
WT: 0.6–1.0 kg

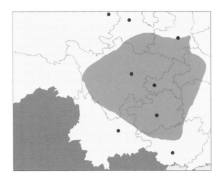

Distribution

The golden pheasant is endemic to China and widely distributed from central to southern China. It has been captive bred for centuries and introduced to Europe and America as game or farm birds. Within the range of this guide, golden pheasants reside across most of Sichuan Province, Guizhou Province, and Chongqing City, with adjoining regions.

Appearance

The golden pheasant has a strong sexual dimorphism, with adult males being unmistakable because of their remarkably long tail and bright plumage. The male's crest and back are bright golden, and the belly is bright red. They have a unique golden nape ruff with black bars, which could be spread like a fan when the male displays in front of females during mating season. The wings of the male are deep blue with a metallic luster. The central tail feathers are elongated and black spotted with cinnamon. Adult females are smaller than males and much less colorful, with light brown plumage with black bars all over the body except for the paler belly. Subadult males are similar to females, but the feathers on the head and back gradually turn colorful with maturity.

Habitat and Ecology

Golden pheasants occur across an elevation range of 1,000–2,500 m in all types of forests and can also be found in shrubs around cultivated lands. They feed on leaves, buds, flowers and seeds, crop seeds, and small invertebrates. The species is polygamous and breeds from March to July. During the mating season, males give loud metallic calls that can be recognized from a far distance. Multiple males gather at open lekking grounds to display their colorful plumage and strut to attract females. They form foraging flocks of several to 20 individuals (up to 80) outside of the breeding season. They run to escape and seldom fly and are difficult to observe because of the thick cover of bamboo thickets or dense shrubs. In central and southern Sichuan Province, where the distribution of the golden pheasant overlaps with the similar Lady Amherst's pheasant, rare hybrid individuals of these two species may occasionally be found in the wild.

Nest and Eggs

Golden pheasants nest on the ground under dense shrubs or in bamboo thickets. The nest is normally a simple shallow pit (20–30 cm wide and 2–8 cm deep) filled with leaf litter and feathers. The clutch size ranges from 5 to 12 eggs, and the eggs are white or light incarnadine without speckles. Eggs are similar in color and size to, although slightly smaller than, farm chicken eggs. Females handle the task of incubation alone, and it normally takes 22–24 days.

Adult male hybrid

Adult female

Juvenile male

Nesting female

Nest

Grey Peacock-Pheasant

Polyplectron bicalcaratum / Hui Kong Que Zhi

Adult male

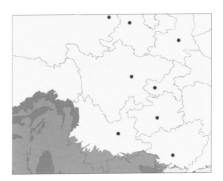

IUCN: LC
CITES: II
CPS: I
CRL: EN

BL: 48–76 cm
TL: 23–40 cm
WT: 0.5–1 kg

Distribution

The grey peacock-pheasant is widely distributed in Southeast Asia (it is the national bird of Myanmar), southern China, and northeast India. Within the range of this guide, grey peacock-pheasants reside in southern Yunnan Province and southern Guangxi AR.

Appearance

The grey peacock-pheasant is a large, grayish-brown pheasant. The male has yellow facial skin and a near-white throat. It has a short, bushy crest on its head. The back and tail of the male are spotted with glossy blue eyespots. The female has an appearance similar to the male but is smaller. The upperparts look scaly, and the eyespots are fewer and indistinct.

Habitat and Ecology

Grey peacock-pheasants inhabit tropical and subtropical forests with dense understory up to 2,000 m elevation. Their diet consists of fruits, seeds, twigs, leaves, worms, and insects. They breed from February to June and appear to be monogamous. The male has a spectacular display, lowering its breast to the ground and uncovering ocelli with open wings and fanned tail. They move solitarily or in pairs. They are wary and hard to observe because of dense understory. When disturbed, they remain still and hide in the understory and will run rather than fly when displaced.

Adult female

Nest and Eggs

The nest is a simple shallow pit filled with leaves and feathers, usually located under shrubs and among grasses. One nest measured was 13 cm wide and 2 cm deep. The clutch size ranges from 1 to 2 eggs, rarely up to 5, and the eggs are light incarnadine without speckles. Only the female incubates the eggs for about 21 days.

Hume's Pheasant
Syrmaticus humiae Hei Jing Chang Wei Zhi

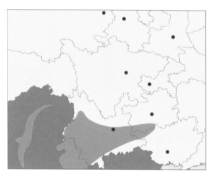

Adult male

IUCN: NT
CITES: I
CPS: I
CRL: VU

BL: 60–90 cm
TL: 20–50 cm
WT: 0.6–1.1 kg

Distribution

Hume's pheasant is native to China, India, Myanmar, and Thailand. Within the range of this guide, Hume's pheasants have been recorded mainly in Yunnan Province, extending into southern Guizhou Province and western Guangxi AR.

Appearance

Hume's pheasant is a large pheasant with marked sexual dimorphism. The male has red facial skin and a metallic black-blue neck. The body plumage is chestnut brown, with a distinctive white band on the scapular and two white wing bars. The lower back and rump is scaled white. It has a long white tail barred with black and brown. The female is all brown with paler buff belly and white-tipped tail. It differs from female Reeves's pheasants in face pattern and shorter tail and from female Elliot's pheasants in white throat and buffier belly.

Habitat and Ecology

Hume's pheasants are found in tropical and temperate forests up to 3,000 m elevation in China. They prefer open forests and forest edges with abundant shrubs or ferns and are most often found in these habitats at 1,000–2,000 m elevation. They feed on plants and insects. The species is polygamous (one male with two females in most cases) and breeds during February and May. Hume's pheasants form forage flocks of three to five individuals during the nonbreeding season. They roost in trees and tall shrubs at night.

Nest and Eggs

The nests are found in densely vegetated areas close to water and forest edges and are about 22–33 cm wide and 5–8 cm deep. The clutch size ranges from 3 to 9 eggs, and the eggs (about 35 × 46 mm) are light incarnadine without speckles. Incubation takes 27–28 days, with only the female sitting on the nest.

Eggs

Female nesting

Adult female

Adult male

Kalij Pheasant

Lophura leucomelanos   Hei Xian

Adult male

IUCN: LC
CITES: III
CPS: II
CRL: NT

BL: 50–75 cm
TL: 19–25 cm
WT: 0.7–1.7 kg

Distribution

The kalij pheasant is distributed main-ly along the Himalayan foothills from Pakistan to western Thailand. It has been introduced to Hawaii. Within the

Adult female

range of our guide it is found close to the country boundary in southern Tibet AR and western Yunnan Province. In the Yingjiang area of western Yunnan Province, kalij pheasants overlap with silver pheasants, and hybrids are commonly recorded.

Appearance

The kalij pheasant is smaller than the silver pheasant and sexually dimorphic. It is mostly black and is called the black pheasant in mandarin. The male has red facial skin and glossy bluish-black body plumage. The color of its crest, rump, and underparts varies among subspecies. Males of the two subspecies in China are black crested. Feathers on their lower back and rump are black and edged in white. The underparts of male *L. l. leucomelanos* are pale grey, and the underparts of *L. l. lathami* are black. The female kalij pheasant is dull brown, with pale-edged fringes to contour feathers reproducing a scaly appearance. The female is also crested and has red facial skin and a pale blue tail.

Habitat and Ecology

Kalij pheasants are found in forested foothills from 245 to 3,700 m elevation. They occupy all types of forests with dense understories and forage in clearings or disturbed areas such as trails, roads, logging areas, and forest edges. They are omnivorous; the bulk of their diet is plant materials, but they also feed on insects and soil invertebrates. The species is monogamous in most cases but can be polygamous. Kalij pheasants breed from April to June and display the same behavior of wing fluttering as the silver pheasant during the breeding season. They are known to hybridize with the silver pheasant where their distributions overlap in northeastern Myanmar and western Yunnan Province.

Nest and Eggs

The nest of the kalij pheasant is a shallow scrape on the ground, lined with leaves and feathers. Nests contain 6–9 eggs, and eggs are creamy white to reddish buff and without spotting. The female incubates the eggs alone, and incubation lasts 20–24 days.

Koklass Pheasant

Pucrasia macrolopha Shao Ji

Adult male

IUCN: LC
CITES: III
CPS: II
CRL: LC

BL: 60–70 cm
TL: 17–22 cm
WT: 0.8–1.5 kg

Distribution

Koklass pheasants are widely distributed in western, northern, and eastern China and exhibit separate populations in the western and central Himalayas. Within the range of our guide, they reside in a band of intermediate elevations from western Yunnan Province into central Shaanxi Province and eastern Chongqing City. There are also populations in northern Guangxi AR.

Appearance

Koklass pheasants are medium-sized pheasants with moderate sexual dimorphism. Adult males have a dark green head, conspicuous white patch on the side of the neck, and remarkable long, trailing ear tufts projecting from the rear of the head. A wide chestnut stripe extends from the throat downward to the vent that can be invisible when seen from the side, and the rest of the body is silver gray and streaked with black. Populations in western and northern China have an orange collar that is absent in southeast populations. Adult females are smaller in size and less colorful, with a much shorter ear tuft. Both sexes have relatively long wedge-shaped tails. Juveniles look similar to females in plumage patterns. Newly hatched chicks can be recognized by their distinctive black eye stripes.

Adult female and chick

Nest

Feathers

Habitat and Ecology

Koklass pheasants inhabit montane forests across a broad elevation range of 600–4,000 m (populations in western and northern China: 1,200–4,600 m; populations in southeast China: 600–1,500 m). They normally forage alone or in pairs, although they can be found in small family groups (three to eight individuals) during winter, and are extremely elusive in dense vegetation. The species is monogamous and breeds from late April to early July. The male makes a loud, harsh territorial call that is distinctive from that of other pheasants and can be heard from long distances. Its common name, the "koklass," is onomatopoeically derived from such calls. The harsh calls are similar to that of male ducks; for that reason koklass pheasants are named "mountain duck" by the local people in southwest China. Males make loud and long calls in the morning before leaving their night roosts, which can be used for population surveys.

Nest and Eggs

Koklass pheasants construct their nests on the ground using a simple shallow pit filled with leaf litter and feathers. The nest is 20–26 cm wide and 8–9 cm deep. The clutch size ranges from 5 to 9, and the eggs are similar to farm chicken eggs but smaller, with reddish brown speckles (size: 4.5 × 3.5 cm). Egg incubation takes 22–27 days and is undertaken solely by females.

Other Sign

The main feathers of this species have black streaks along the main axis that make them distinct from other sympatric pheasants.

Lady Amherst's Pheasant

Chrysolophus amherstiae　Bai Fu Jin Ji

Adult male

IUCN: LC
CITES: NL
CPS: II
CRL: NT

BL: 60–180 cm
TL: 28–110 cm
WT: 600–900 g

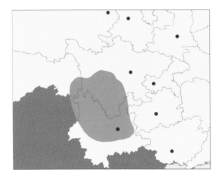

Distribution

Lady Amherst's pheasants are distributed in southeast China and northern Myanmar, and wild populations have become established in England following translocation. Within the range of this guide, they reside in southeast Tibet AR and southwest Sichuan and Yunnan Provinces. They overlap with their close relative, the golden pheasant, through central and southern Sichuan Province.

Appearance

The Lady Amherst's pheasant is a medium-sized pheasant that is sexually dimorphic. The male has an extremely long tail and colorful plumage. It has a white neck ruff with black scaling, and the throat, breast, and upperparts are metallic dark green. The male has a short, purple-red crest atop the head that can be erected during the mating display. It has dark blue wings and a white belly. The tail can be up to 100 cm long, white with thin black barring, and the upper tail coverts are tipped with orange red. The tail feathers of the birds are used for head decorations by the traditional Beijing opera. Adult females look similar to female golden pheasants but have warmer plumage tones and are more boldly barred. Their bills and legs are gray, whereas those of female golden pheasants are yellow. The crown of the female is chestnut brown, contrasting with its paler face and throat. The Lady Amherst's pheasant is slightly larger than the closely related golden pheasant. Hybrids of the two species are found in captivity and in the wild, as early as depicted in a painting by Emperor Huizong of the Song dynasty (12th century).

Habitat and Ecology

Lady Amherst's pheasants inhabit forests from 1,500 to 3,600 m elevation. They occur primarily in conifer-broadleaf mixed forest and broadleaf forest with dense understory and are less common in conifer forest and shrubs.

They feed on grain, leaves, and invertebrates. The species is polygamous (one male with two to four females) and breeds from April to June. The male spreads its ruff feathers like a fan during the courtship display, a behavior similar to that of the male golden pheasant. They run when startled and can suddenly burst upward with heavy wingbeats.

Nest and Eggs

The female constructs a simple nest on the ground filled with leaves and feathers under dense shrubs and in bamboo thickets. The nest is usually 25 × 30 cm in size and 5–13 cm deep. Its clutch size ranges from 4 to 9 eggs, which are creamy white or light incarnadine without speckles. Only the female incubates the eggs, and incubation normally takes 21–23 days.

Adult female

Juvenile male

Red Junglefowl

Gallus gallus　Hong Yuan Ji

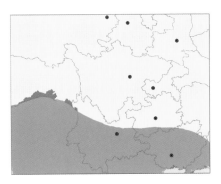

Adult and subadult males

IUCN: LC
CITES: NL
CPS: II
CRL: NT

BL: 40–80 cm
TL: 14–40 cm
WT: 0.5–1.1 kg

Distribution

The red junglefowl is native to South and Southeast Asia, and its large range extends from northern and eastern India eastward across southern China and into Indonesia and the Philippines. Within the range of this guide, red junglefowl reside in Yunnan Province and Guangxi AR.

Appearance

The red junglefowl is a relatively large pheasant with sexual dimorphism. It is similar to the domestic chicken (which is considered the domesticated form of this species) but longer and thinner. The male has a red comb and wattles, covered with golden to bronze hackles over the rear crown, neck, and mantle. The tails are metallic green black with two central tail feathers elongated and sickle shaped. The female is much smaller than the male and is all brown and streaked with buff. The neck feathers of the female are slightly elongated and edged with straw yellow. The male red junglefowl has a long, sharp spur on its lower leg, whereas the female lacks spurs.

Habitat and Ecology

Red junglefowl occur in a wide range of habitats in tropical areas, including scrubs, open woodlands, forests, grasslands, and farmlands below 1,850 m elevation. They feed on grains, fruits, roots, termites, ants, and other insects. The species is polygamous and breeds from February to June. Red junglefowl form family flocks of six to seven individuals (up to 20), typically one male accompanied by several females and chicks, outside of the breeding season. They are wary, with good sight and hearing, ever ready to flush when disturbed. When startled, they quickly run or fly into cover. The wild populations of this species are faced with threats of hybridization with free-ranging domestic chickens.

Nest and Eggs

The red junglefowl constructs nests on bare ground below bushes close to trail and forest edges. The nest is about 19–22 × 17–23 cm wide and 2–5 cm deep. Its clutch size ranges from 3 to 7, and the eggs are smooth without speckles, very similar to domestic chicken eggs. The female is responsible for incubation, which takes about 23 days.

Reeves's Pheasant

Syrmaticus reevesii // Bai Guan Chang Wei Zhi

Adult male

IUCN: VU
CITES: NL
CPS: II
CRL: EN

BL: 150–210 cm
TL: 25–150 cm
WT: 0.7–1.7 kg

Distribution

Reeves's pheasant is endemic to central and southern China. Within the range of this guide, it resides in eastern Sichuan

Province through Chongqing City and northern Guizhou Province.

Appearance

Reeves's pheasant is the largest among the four *Syrmaticus* species found in China. The species is sexually dimorphic, with the male remarkably colorful. It has a white crown, a white patch under the eye, and a white collar, and the rest of the head is black. The upperparts are scaled golden yellow, whereas the underparts are white scaled with both black and chestnut. It has an extremely long white tail edged with chestnut and barred with black. The tail can be as long as 1.5 m, and the tail feathers are used for head decorations by the traditional Beijing opera. The female is much smaller and less striking. It has a brown and streaked body, a pale buff face with a distinctive brown crown and cheek bar, and a much longer and pointing tail than the females of the other three *Syrmaticus* species.

Habitat and Ecology

Reeves's pheasants are found throughout forest habitats across an elevation between 200 and 2,600 m. They inhabit both primary and secondary forests and also occur in farmlands close to forest edges. They feed primarily on plant materials, including crop seeds during the winter. They breed from March to June. The breeding associations are unknown, but mated pairs are observed more often than other gender combinations. The male is frequently observed during the mating season to perform wing fluttering to claim its territory and court the female. Small feeding flocks are observed outside of the breeding season.

Nest and Eggs

The nest is a shallow pit on the ground filled with leaves, twigs, and feathers. It is often covered by high grass or shrubs and is close to trails. The nest is about 18–24 cm wide and 5–9 cm deep. The clutch size ranges from 6 to 14 eggs, which are light incarnadine without speckles. Only the female incubates the eggs, which takes 26–27 days.

Adult female

Nest

Silver Pheasant

Lophura nycthemera   Bai Xian

Adult male

IUCN: LC	BL: 70–120 cm
CITES: NL	TL: 28–80 cm
CPS: II	WT: 1.1–2.0 kg
CRL: LC	

Distribution

The silver pheasant is widely distributed in the provinces of eastern and southern China and extends its distribution southward into the Indochina peninsula. Within our geographic range, it resides in central and southern Yunnan Provinces and Guangxi AR, with isolated populations in Sichuan and Guizhou Provinces.

Appearance

The silver pheasant is a relatively large pheasant with obvious sexual dimorphism. The male has extensive red facial skin and a bluish-black crest. Contrasting with its dark underparts, the upperparts and the tail are white, closely marked with thin V-shaped black stripes. The females are brown, with a much shorter tail. The legs of the male are red, distinguishing them from male kalij pheasants, whose legs are gray. The subspecies have gradual variations in plumage color, and both sexes are variable. The upperparts and outer tail feathers of the male in the south are more heavily vermiculated, making them appear darker than the northeast subspecies. Females are typically plain brown, although some races have white

Male (back), subadult male (middle), and a female (front)

Wing fluttering

Nest

streaks on the belly or black-and-white vermiculation on the outer tail feathers. The male juveniles are brown, gradually changing to adult male plumage in their second year, so the intermediate form has mixed white and brown patches of feathers.

Habitat and Ecology

Silver pheasants are found in subtropical and tropical forests, shrubs, and open habitats near forest edges in low mountains and hills at elevations up to 2,000 m. They feed on a variety of plant materials and insects. The species is polygamous and breeds from April to June. The male performs wing fluttering to court the female during the breeding season. Wing fluttering consists of 8–10 rapid strokes of half-extended wings while the body is in an upright posture. The birds form foraging flocks of 3–10 individuals during the nonbreeding season. Their calls are seldom heard in the wild. When disturbed, they make sharp prolonged whistles to warn the flock before all birds quickly run or fly away. Flushed birds make low volume calls, "gu-gu-gu," which appear to help them rejoin. They roost in big trees close to ravines at night.

Nest and Eggs

The nest is a shallow scrape on the ground sheltered under a rock or around dense bushes and is made of dry leaves and feathers. The clutch ranges from 4 to 9 eggs, with eggs appearing light incarnadine without spotting and similar in color and size to farm chicken eggs. Incubation, which lasts 24–25 days, is carried out solely by the female.

Other Sign

Loose feathers of the silver pheasant, which are white with thin black stripes, are easily recognized, especially the long flight and tail feathers.

Tibetan Eared Pheasant
Crossoptilon harmani // Zang Ma Ji

Adult

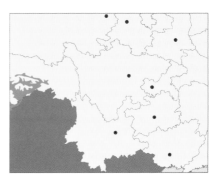

IUCN: NT
CITES: I
CPS: II
CRL: NT

BL: 75–85 cm
TL: 45–56 cm
WT: 1.5–2.5 kg

Distribution

The Tibetan eared pheasant was formerly considered a subspecies of the white eared pheasant. Within the range of our guide the pheasant is found in Lhasa City and southern Tibet AR.

Appearance

The Tibetan eared pheasant has grayish-blue plumage and is sexually monomorphic. It looks similar to the blue eared pheasant, but their ranges do not overlap. In contrast to the blue eared pheasant, the Tibetan eared pheasant has short ear tufts that do not extend beyond the back of its head. It has a white chin and belly and light gray rump and upper tail coverts but lacks the white base on the outer tail feathers.

Habitat and Ecology

Tibetan eared pheasants occur in primary forests, alpine shrubs, and meadows from 3,000 to 5,000 m elevation.

They forage in open meadows, especially along river valleys where the soil is moist, making it easier for them to dig for food. The species is monogamous, but birds commonly forage in flocks of 15–30 individuals outside of the breeding season. A large flock can have >100 individuals during winter. They roost in big trees and dense shrubs, preferably close to a rock cliff, and use the same roost site year-round. Hybrid individuals of the Tibetan eared pheasant and white eared pheasant are observed where the two species overlap in the Brahmaputra River Valley, Tibet AR.

Female

Nest and Eggs

The nest is a shallow scrape, lined with dry leaf litter and feathers, under big trees and shrubs and in rock cavities. Nests are normally 31 cm (27–39 cm) in diameter and 6.8 cm (3.2–10.6 cm) in depth. The clutch size ranges from 4 to 11 eggs, and the eggs are white, light brown, or light green, with or without brown speckles. Only females incubate the eggs, and incubation lasts 24–25 days.

Nest

White Eared Pheasant
Crossoptilon crossoptilon　Bai Ma Ji

Adult

IUCN: NT	BL: 80–110 cm
CITES: I	TL: 40–53 cm
CPS: II	WT: 1.2–3.0 kg
CRL: NT	

Distribution

The white eared pheasant has its distribution in southwest China and is found entirely within the range of our guide. It resides in mountain valleys in western Sichuan, southeast Qinghai, and northwest Yunnan Provinces and eastern Tibet AR.

Appearance

The white eared pheasant is the largest *Crossoptilon* pheasant and is sexually monomorphic. It is predominantly white; some have washed gray on their upperparts. Its tail feathers and wingtips are black, and the legs are red. It has a black cap, red bare facial skin, and white ear tufts. The ear tufts, rather soft body feathers, and long, loose, arched tails are characteristics of all *Crossoptilon* pheasants.

Habitat and Ecology

White eared pheasants occur in coniferous and mixed forests, alpine shrubs, and alpine meadows between 3,000 and 4,600 m elevation. They are often seen foraging in open meadow along forest edges in river valleys or above tree line. Their diet consists mainly of plant parts. The species forms monogamous pairs in spring, and the male and female accompany each other throughout the breeding season. They live in flocks of up to 30 individuals outside of the breeding season; a large flock of 200 individuals was reported during winter in Daocheng County, Sichuan Province. Cooperative young-caring behavior has been observed; mixed broods form a social unit, and nonbreeding subadults or adults may help the parents to fledge young. A flock roosts together in big trees at night. The species benefits from the cultural protection of local Tibetans and can be fairly common around Tibetan monasteries.

Nest and Eggs

The nest is a simple, shallow pit on the ground under shrubs or rock overhangs, 26–34 cm in diameter and 4–10 cm in depth. The clutch size ranges from 5 to 11 eggs, and the eggs are pale stone without speckles. Incubation takes 22–28 days and is conducted solely by the female. Chicks are cared for by the mated pair and juvenile birds.

Nest

Female with chicks

"Kindergarten"

PARTRIDGES / CHUN

Buff-Throated Partridge
Tetraophasis szechenyii　Huang Hou Zhi Chun
Szechenyi's Monal Partridge

Adult male

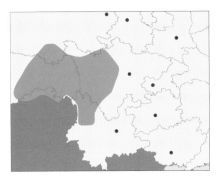

IUCN: LC
CITES: NL
CPS: NL
CRL: VU

BL: 40–55 cm
TL: 17–19 cm
WT: 0.9–1.8 kg

Distribution

The buff-throated partridge is distributed on the Qinghai-Tibet Plateau and the eastern Himalayas in China and extreme northeast India. Within the range of this guide, buff-throated partridges reside in eastern Tibet AR, southern Qinghai Province, and western Sichuan and northern Yunnan Provinces.

Appearance

Buff-throated partridges are large partridges that do not exhibit sexual dimorphism in plumage, but males are usually larger than females. They have an orange-buff throat and dark gray breast with black streaks. Buff to whitish bandings appear on their scapulars

and wings. The flanks are gray with large rusty streaks. They have a comparatively long and white-tipped tail. The orbital skin is bright red in adults but paler in juveniles. They look similar to their close relative, the chestnut-throated partridge, but chestnut-throated partridges have a white margin around their throat.

Habitat and Ecology

Buff-throated partridges inhabit subalpine forests (conifer forest and alpine oak forest), shrub forest (common in rhododendron stands), and alpine meadows across an elevation range of 3,300–4,600 m. They use their strong bill, rather than claws, to dig for food. They feed on plant tubers, roots, young leaves, seeds, berries, and insects. The species is monogamous and breeds from March to July. They forage in small family groups (group size typically three to five individuals) outside of the breeding season. They are typically elusive but can be habituated to humans around Tibetan monasteries in region.

Nest and Eggs

Buff-throated partridges nest both in the trees (spruce, fir and oak) and on the ground among rhododendron or oak shrubs. The nest is made of dry leaf litter, moss, lichens, and downy feathers, approximately 27 × 29 cm wide and 4–6 cm deep. The clutch size ranges from 3 to 7 eggs, and the eggs are light brown with thick dark brown speckles (size 5 × 3.5 cm). Incubation takes 25–30 days and is undertaken solely by females. Offspring from a previous season may stay with their parents and assist in the care of chicks.

Male and female

Adult female (left) and juvenile (right)

Nest

Chestnut-Throated Partridge

Tetraophasis obscurus / Hong Hou Zhi Chun

Verreaux's Monal Partridge

Adult

IUCN: LC
CITES: NL
CPS: I
CRL: VU

BL: 44-55 cm
TL: 15-18 cm
WT: 0.9-1.8 kg

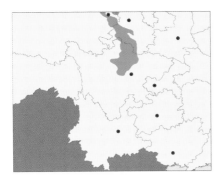

Distribution

Chestnut-throated partridges are endemic to the mountains of central China. Within the range of this guide, they reside in the mountains of northern Sichuan Province through western Gansu and eastern Qinghai Provinces.

Appearance

The chestnut-throated partridge is a large partridge without marked sexual dimorphism, although males are slightly larger than the females. It looks similar to the buff-throated partridge, *T. szechenyii*. The main difference between the two species is the coloration of the throat, as described in their English names. In addition, the chestnut-throated partridge has a white ring around its throat that is lacking in the buff-throated partridge. Although there was concern that *T. obscurus* and *T. szechenyii* are two color morphs of the same species, recent genetic studies indicate they are distinct species.

Habitat and Ecology

Chestnut-throated partridges occur in coniferous forests, in rhododendron and juniper shrubs, and on open rocky slopes spanning the tree line and above (3,000 to 4,880 m elevation). They feed on plant material and insects and especially favor roots and tubers. The species is thought to be monogamous and breeds from April to July, and the females lay their eggs in May. They form foraging groups of three to five individuals outside of the breeding season. Local people believe they act as weather forecasters because they make harsh calls before rainfall. They fly downhill into forest cover when disturbed.

Nest and Eggs

Chestnut-throated partridges nest both in the trees (rhododendron and oak) and on the ground among shrubs. The clutch size ranges from 3 to 7 eggs, and the eggs are light brown with thick dark brown speckles. Incubation takes 27–28 days and is undertaken solely by the females.

Adult

Chinese Bamboo Partridge

Bambusicola thoracicus | Hui Xiong Zhu Ji

Adult pair

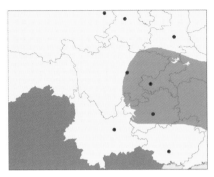

IUCN: LC
CITES: NL
CPS: NL
CRL: LC

BL: 27–38 cm
TL: 9–11 cm
WT: 200–350 g

Distribution

The Chinese bamboo partridge is endemic to central and eastern China and has been introduced into Japan and Hawaii. Within the range of our guide, Chinese bamboo partridges reside in a region east of Chengdu (Sichuan Province) bordered in the north by Shaanxi Province and in the south by Guangxi AR.

Appearance

The Chinese bamboo partridge is a small pheasant with no sexual dimorphism. It has a distinctive face pattern with a long bluish-gray supercilium that runs from the forehead down to the neck and a contrasting rufous color below the eye and around the throat. The breast is bluish gray, bordered below by a strip of rufous. The upperparts are gray, boldly marked with large chestnut spots. The underparts down from the breast are buff with heart-shaped black spots on the flank.

Habitat and Ecology

Chinese bamboo partridges inhabit dry grasslands and shrublands to low hills up to 2,000 m elevation but mostly below 1,000 m. They prefer bamboo stands but are not entirely dependent on bamboo. The species is highly tolerant to disturbance and can be found in small urban parks. When they forage in bamboo stands, they are readily audible as they step on dry leaves and search for food. When disturbed, they hide in thick cover and are readily flushed if closely approached. They make heavy wingbeats while flushing and fly brief distances and then quickly run away. Their loud, distinctive calls of "ke-pu-kwai" are repeated often and can be heard throughout the year. They feed on leaves, buds, seeds, grains, and invertebrates. The species appears to be monogamous and breeds from March to August. They form flocks of 2–20 individuals outside of the breeding season. They roost in bamboo or trees at night.

Nest and Eggs

The Chinese bamboo partridge nests on the ground at the base of a tree, under shrubs, or in the grass. The nest is a shallow pit made of leaves, grasses, and downy feathers, about 10–15 cm in diameter and 2–4 cm in depth. The clutch size ranges from 6 to 9 eggs, which are creamy or light brown with tiny brown speckles, similar in size to quail eggs. Females incubate the eggs alone, and incubation usually lasts 16–18 days (up to 20 days).

Adult pair

Mountain Bamboo Partridge
Bambusicola fytchii Zong Xiong Zhu Ji

Adult pair

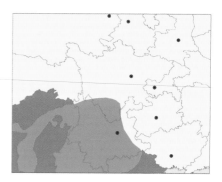

IUCN: LC
CITES: NL
CPS: NL
CRL: LC

BL: 27–38 cm
TL: 9–13 cm
WT: 250–400 g

Distribution

The mountain bamboo partridge is distributed in the foothills of the eastern Himalayas in southern Asia. Within the range of this guide, it resides in Yunnan Province and southwest Sichuan Prov-

ince. The range shown is increased from that indicated by IUCN on the basis of our local knowledge.

Appearance

The mountain bamboo partridge is a small pheasant, with slight sexual dimorphism, as the male is slightly larger than the female and has spurs. It has a distinctive black eye stripe running from the eye down to the neck. The forehead and supercilium are buff white. The throat and breast are buff, heavily streaked with dark rufous on the breast. The dorsal plumage is gray with large brown to black speckles. It has a creamy-white belly marked with dense heart-shaped black spots. The tail is rufous brown and barred with buff white.

Habitat and Ecology

The mountain bamboo partridge occurs in tall grassland, open scrub and woodland, and bamboo patches, often close to water, up to 2,000 m elevation. Its diet consists of leaves, buds, seeds, berries, roots, and invertebrates. The species is monogamous and breeds from April to July. Small coveys of five to six family members are found outside of the breeding season. When startled, they fly only a few meters before landing in dense cover.

Nest and Eggs

The nests of mountain bamboo partridge are a simple scrape on the ground amid grass, scrub, or bamboo. The nest is about 16 cm in diameter and 2–4 cm in depth. The clutch size averages 4–5 (range 3–7), and eggs are creamy buff to deep buff and without speckles. The female incubates the eggs alone, which lasts about 20 days.

Przewalski's Partridge
Alectoris magna　Da Shi Ji

Adult

IUCN: LC
CITES: NL
CPS: NL
CRL: NT

BL: 36–38 cm
TL: 9–13 cm
WT: 400–700 g

Distribution

Przewalski's partridge is endemic to the arid mountains of central China. Within the range of this guide, it resides in eastern Qinghai and southern Gansu Provinces.

Appearance

Przewalski's partridge is a small partridge, with little sexual dimorphism. It has a red bill and feet. The chin, throat, and upper chest are white, bordered by a rusty brown collar and a black inner line. The upperparts are sandy gray, with a slight brown wash on its back and wings. The flanks are white barred with black and chestnut. Przewalski's partridge is one of two species in the genus *Alectoris* in China, along with the chukar (*Alectoris chukar*), which is smaller and has a black collar. The two species' ranges meet in Gansu Province adjacent to Qinghai Province, and hybrids are found in the Liupan Shan Mountains.

Habitat and Ecology

Przewalski's partridges occur on open rocky or grassy hills scattered with small bushes in arid and semiarid regions between 1,800 and 2,500 m elevation. They descend to lower elevations in valleys during winter and can be found foraging near houses. They feed on leaves, seedlings, seeds, tubers and roots, and a small proportion of insects. The species is monogamous and breeds from late March to July. Unmated males form small coveys during the breeding season and winter flocks of mixed sex can be up to 30 individuals. They roost on the ground in rock crevices.

Nest and Eggs

The nest of Przewalski's partridge is a simple scrape on the ground, lined with vegetation and downy feathers. The clutch size ranges from 7 to 20 eggs, and the eggs are pale yellow buff or pale olive brown with brown speckles. The female incubates the eggs alone, and the eggs hatch after 22–24 days. The male stays with the female and assists in the care of the chicks.

Snow Partridge

Lerwa lerwa Xue Chun

Adult

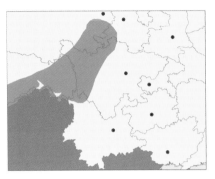

IUCN: LC
CITES: NL
CPS: II
CRL: NT

BL: 30–40 cm
TL: 17–19 cm
WT: 400–700 g

Distribution

The range of the snow partridge extends from Pakistan eastward through the Himalayas into the Qinghai-Tibet Plateau and central China. Within the range of this guide, they reside in southern Gansu, eastern Qinghai, western Sichuan, and northwest Yunnan Provinces, and southern Tibet AR.

Appearance

The snow partridge is a small partridge with minimal sexual dimorphism. The bill and feet of the bird are red. The head, tail, and entire upperparts are finely barred in black and white, whereas the underparts are white streaked with rich dark chestnut. The female is slightly smaller than the male, is duller in color (pinkish bill), and lacks spurs.

Habitat and Ecology

Snow partridges occur in rocky and grassy slopes interspersed with scrubs

and scree between 2,900 and 5,000 m elevation. They feed on leaves, buds, shoots, and seeds from grasses and shrubs, as well as moss, lichens, and insects. They breed from April to July. The species is monogamous, and the male helps to fledge young. They are usually encountered in pairs or small coveys (6–8 individuals but can be 20–30) outside of the breeding season.

Nest and Eggs

The snow partridge builds nests under rock overhangs or in dense shrubs or grass. The clutch size ranges from 3 to 5 eggs, and the eggs are pale buff with fine reddish-brown speckles.

Flock

Tibetan Partridge

Perdix hodgsoniae | Gao Yuan Shan Chun

Adult

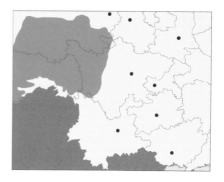

IUCN: LC	BL: 27–32 cm
CITES: NL	TL: 7–10 cm
CPS: NL	WT: 200–500 g
CRL: LC	

Distribution

The Tibetan partridge is widely found throughout the Tibetan Plateau. Within the range of this guide, they reside in extreme southwest Gansu, southern Qinghai, and western Sichuan Provinces, as well as Tibet AR.

Appearance

Tibetan partridges are small pheasants and exhibit no sexual dimorphism. They have a striking facial pattern different from other Perdix species, with a white forehead connecting with the long white supercilium. The chin, face, and throat are white; ear coverts are gray to black, and a black strip runs down the face from below the eye. They have a rufous collar. The upperparts are grayish brown, finely streaked with dark brown spotting and white shaft. The underparts are white, barred with black on the belly and black and chestnut on the flank.

Habitat and Ecology

Tibetan partridges inhabit rocky slopes and alpine meadows with stunted trees and scrubs between 2,800 and 5,200 m elevation. They forage on the ground in sparsely vegetated areas and avoid dense rhododendron and oak stands. They are also found in farmlands and rangelands along the river valleys in eastern Tibet AR. They feed on leaves, shoots, flowers, seeds, moss, and small invertebrates. The species is monogamous and breeds from late March to July, sometimes August. They can form foraging groups up to 30 individuals (typically 5–12) during the nonbreeding season. When disturbed, they seldom take flight but run and occasionally burst into the air and scatter in all directions with shrill calls. They roost in shallow pits on the ground under dense shrubs, and a single pit may be shared by multiple individuals.

Nest and Eggs

The nests of Tibetan partridges are constructed on the ground under shrubs, trees, or rock overhangs, approximately 18 cm in diameter (15–24 cm) and 7 cm (5–9 cm) in depth. The clutch size ranges from 5 to 12 (average of 8) eggs. The eggs are pale in color and have no speckles (size 4 × 2.8 cm). Incubation takes about 23 days and is undertaken solely by the females. Both the male and the female tend the chicks.

Nest

Adult

Flock

TRAGOPANS, MONALS, AND OTHERS
JIAO ZHI, HONG ZHI

Blyth's Tragopan
Tragopan blythii Hui Fu Jiao Zhi
Grey-Bellied Tragopan

Adult male

IUCN: VU
CITES: I
CPS: I
CRL: DD

BL: 53–70 cm
TL: 17–22 cm
WT: 0.9–1.6 kg

Distribution

Blyth's tragopan is distributed in the eastern Himalayas and Indian subcontinent from southeast China through northern Myanmar to eastern Bhutan and northeast India. Within the range of this guide, it resides in southeast Tibet AR and northwest Yunnan Province.

Appearance

Blyth's tragopan is a medium-sized pheasant that is sexually dimorphic. The male has orange-yellow facial skin, two inflatable blue horns, and a blue-edged yellow lappet. The neck and breast are deep orange red. The upperpart is brown, barred with black and mottled with black-rimmed white spots. The underpart is gray from the lower breast downward. The female looks similar to the female Temminck's tragopan but has yellow orbital skin (blue in Temminck's) and lacks bold white spots on the belly.

Habitat and Ecology

Blyth's tragopans inhabit various forests, including alpine oak, rhododendron, and temperate to subtropical forests with dense undergrowth. They primarily occur between 1,800 and 2,400 m elevation but occasionally are reported up to 3,300 m elevation during summer. They are generalists with a diet including berries, leaves, buds, seeds, and small invertebrates. The mating and social systems of the species are poorly known.

Satyr Tragopan

Tragopan satyra Hong Xiong Jiao Zhi

Adult male

IUCN: NT	BL: 57–72 cm
CITES: III	TL: 19–29 cm
CPS: I	WT: 1–2.1 kg
CRL: VU	

Distribution

Satyr tragopans have a narrow distribution along the southern side of the central Himalayas in China, Bhutan, Nepal, and India. Within our geographic range, they reside in a limited area of southern Tibet AR.

Appearance

The Satyr tragopan is a medium-sized pheasant that is sexually dimorphic. The adult male is bright crimson from its neck through its breast to the belly. It has a black head, blue facial skin and lappet, and an orange stripe on crown sides. The upperparts are rufous brown to grayish brown. Both its upper- and underparts are covered with conspicuous black-bordered white spots that become progressively larger on the belly toward the vent. The adult female is rufous brown to grayish brown, finely mottled with white shaft streaks.

Habitat and Ecology

Satyr tragopans inhabit moist alpine oak and rhododendron forests between 2,200 and 4,250 m elevation. They may move down to elevations as low as 1,800 m during the winter. This species may prefer forests with dense understory, but their ecology and social organization is poorly known. Like other tragopan species, the male performs a magnificent display during courtship, showing off its erected horns and inflated gular lappet and whirring its wings in an arched and upright posture.

Temminck's Tragopan

Tragopan temminckii   Hong Fu Jiao Zhi

Adult male

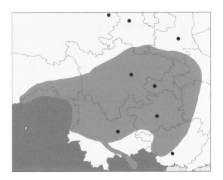

IUCN: LC
CITES: NL
CPS: II
CRL: NT

BL: 44–60 cm
TL: 15–24 cm
WT: 0.9–1.6 kg

Distribution

Temminck's tragopan is mainly distributed in southwest China and also occurs in northeast Assam (India), northern Myanmar, and northwest Vietnam. Within the range of our guide, it resides in all provinces and ARs east of the plateau.

Appearance

Temminck's tragopan is a medium-sized pheasant that is sexually dimorphic. The adult male is bright orange crimson with upperparts covered in black-rimmed white spots and underparts covered in pale gray spots. The head of the male is black with orange stripes on its crown sides, and the facial skin and lappet are blue. The male inflates its horns and lappet during courtship, creating a striking blue pattern with contrasting streaks of red. The female is grayish brown with bold white underpart spotting and white shaft streaks all over the body.

Habitat and Ecology

Temminck's tragopans inhabit broadleaf and mixed broadleaf-conifer forests between 1,000 and 3,500 m elevation. They prefer steep forest slopes with dense understory and are also found in bamboo forest and rhododendron shrubs. They forage in grassy areas, preferably on southern slopes, and rest in bamboo and shrubs in the sun after feeding. Plants constitute the bulk of their diet and include a range of seeds, fruits, flowers, and leaves. The mating system is unknown, appearing to be monogamous, but two females accompanying one male is occasionally observed during the breeding season (late March–July). They form foraging flocks of two to five individuals during winter. They roost at 3–10 m height in large trees and occasionally in bamboo forest and rhododendron shrubs.

Nest and Eggs

Temminck's tragopans construct nests in trees that can be up to 8 m above the ground. The nest is made of twigs, leaves, moss, and feathers. The clutch size ranges from 3 to 5 eggs. Eggs are buff with brown speckles. Incubation, which takes 25–27 days, is solely conducted by the female.

Other Sign

Where a Temminck's tragopan has been predated, it is common to find the contour feathers of the bird, gray with a crimson tip, scattered on the ground.

Adult female

Male display

Subadult male (front) and adult male (back)

Nest

Male feathers

Chinese Monal

Lophophorus lhuysii   Lv Wei Hong Zhi

Adult male

IUCN: VU
CITES: I
CPS: I
CRL: EN

BL: 72–80 cm
TL: 27–30 cm
WT: 3–4 kg

Distribution

The Chinese monal is endemic to southwest China. Within the range of our guide, Chinese monals reside in western Sichuan Province and adjoining regions.

Appearance

Chinese monals are large pheasants with marked sexual dimorphism. The male is highly iridescent and unmistakable. It has a green head, copper nape, bluish green upperparts, and a long purple crest. The lower back of the male is white and visible during flight. The female is dark brown, finely barred with rufous brown, and streaked with buff shafts. The throat and back of the female are white.

Habitat and Ecology

Chinese monals are found in subalpine and alpine shrubs and meadows with exposed rocks and cliffs above the tree line at 3,000–5,000 m elevation. They may descend to 2,000 m elevation into forests during winter. They dig roots and tubers for food and especially favor the tubers of the genus *Fritallaria*; they are named "Beimu Ji" ("Fritallaria pheasant" in Chinese) by some local people. They breed from late March to June, and the breeding system is unclear. The males have been observed making display flights, gliding horizontally but slowly downward like an eagle; the species is also called "Ying Ji" ("eagle pheasant" in Chinese) locally. The birds stay in small flocks of one to eight individuals during the nonbreeding season. They roost in trees or on the ground with cliffs or shrubs for cover.

Droppings

Adult female

Nest and Eggs

Chinese monals construct their nests on the ground under dense shrubs or in rock crevices. The nest is a shallow pit made of moss, grass, and feathers, approximately 20–36 cm wide. The clutch size ranges from 3 to 4 eggs, and eggs are yellowish brown with brown speckles (size 6.9 × 4.6 cm). Incubation is undertaken solely by the females and lasts about 28 days.

Other Sign

The droppings of all three monal species are easy to recognize because of their large size (diameter >2 cm, much larger than that of other sympatric pheasants) and the unique habitat in which they are located (i.e., alpine meadows, shrubs, and forest at or above the tree line). Monals dig roots and tubers for food and leave holes (3–10 cm wide, 8–12 cm deep) in the ground after feeding.

Himalayan Monal

Lophophorus impejanus   Zong Wei Hong Zhi

Male in display and female

IUCN: LC
CITES: I
CPS: I
CRL: NT

BL: 70 cm
TL: 21–24 cm
WT: 1.8–2.4 kg

also been recorded in Myanmar and is the national bird of Nepal. Within the range of our guide, Himalayan monals reside in southern Tibet AR. There was one report from Gongshan County, northwest Yunnan Province.

Appearance

The Himalayan monal is smaller than the Chinese monal, but otherwise, the two species are similar in appearance. The notable difference between the two is that the male Himalayan monal has a green (not purple) crest and a cinnamon-brown (not bluish-green) tail. The female Himalayan monal has white upper tail coverts (back is not white) and a narrow white strip on the tip of the tail. The male lowers its head, droops its wings, and fans its tail during courtship, exhibiting its magnificent colors in front of the female.

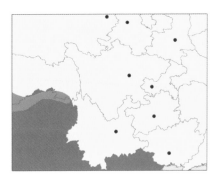

Distribution

The Himalayan monal has a large range, extending from eastern Afghanistan to Bhutan through the Himalayas. It has

Habitat and Ecology

Himalayan monals favor rocky and grass-covered slopes interspersed with scrubs. They reside above the tree line but move down into birch, oak, rhododendron, and conifer forests during winter. They are found at elevations of 3,000–4,100 m in southern Tibet AR. Roots and tubers are their most important food items, although they also feed on leaves, seeds, and invertebrates. They forage in small groups of up to four individuals and split into pairs during the breeding season from April to June. They roost on the ground or in rock crevices and in trees.

Nest and Eggs

Himalayan monals construct their nests on the ground under the shelter of rocks or shrubs. The nest is a shallow pit made of moss, grass, leaves, and feathers. The clutch size ranges from 4 to 6. The eggs are pale yellowish or reddish buff with brown speckles (size 6.4 × 4.5 cm). Incubation is undertaken solely by the females.

Adult female

Sclater's Monal

Lophophorus sclateri   Bai Wei Shao Hong Zhi

Male

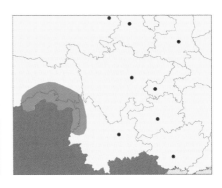

IUCN: VU
CITES: I
CPS: I
CRL: EN

BL: 56–68 cm
TL: 18–22 cm
WT: 2–3 kg

Distribution

Sclater's monal is distributed in the eastern Himalayas and the Hengduan Mountains. Within our geographic range, Sclater's monals reside in western Yunnan Province and southeast Tibet AR.

Appearance

Similar to the other two monal pheasants, the adult male Sclater's monal has multicolor plumage throughout, whereas the female is dull in color. The male lacks the crest of both the Chinese monal and Himalayan monal and has a rufous tail with a white tip. It has a much broader white patch on the back than the other two monal pheasants. The lower back, rump, and upper tail coverts of the female Sclater's monal are light brown, in contrast with the dark brown on its upper body.

Habitat and Ecology

Sclater's monals inhabit alpine and subalpine forests, rhododendron shrubs, and grasslands and are frequently found foraging in bamboo forests, open lands interspersed with scrubs, or along forest edges. They occur within an elevation range of 2,500–4,200 m. As with the other two monal species, Sclater's monals can tolerate snow and harsh weather but may move to lower-elevation forests during winter. They breed from late March to June, and the breeding system is unknown. They roost under rocks or in shrubs on steep slopes.

Nest and Eggs

The nest of Sclater's monal is a shallow pit constructed with grass and feathers (20–30 cm diameter, depth of 8 cm). The clutch size ranges from 2 to 3. The eggs are dark yellowish with irregular brown speckles (7 × 4.7 cm). Incubation takes 28 days and is undertaken solely by the females.

Nest

Adult female

Chinese Francolin

Francolinus pintadeanus   Zhong Hua Zhe Gu

Adult male

IUCN: LC
CITES: NL
CPS: NL
CRL: NT

BL: 30-34 cm
TL: 9-11 cm
WT: 280-400 g

Distribution

The Chinese francolin is distributed from eastern and southern China, extending southward into India and Southeast Asia. It is the only species within the genus *Francolinus* in China. Within the range of this guide, it resides in southern Yunnan Province and Guangxi AR.

Appearance

The Chinese francolin is a small pheasant with slight sexual dimorphism. The male has a distinctive pace pattern, with a brown crown and nape, black eye stripes, and a rufous supercilium. A black moustache separates its white cheek from the white chin and throat. The upperparts are black and heavily mottled with white spots on the mantle that gradually turn into white barring on the rump and tail. With broader white fringes, the lower underparts

appear whiter than the upperparts. The female is slightly smaller than the male and looks duller.

Habitat and Ecology

Chinese francolins inhabit dry, open forest, grassland, and shrubland up to 1,600 m elevation. They are shy and

Female

wary, moving under vegetation most of the time. Their diet consists of leaves, seeds, flowers, roots, and insects. The species is monogamous and breeds from March to June. They are usually observed in pairs or family groups.

Nest and Eggs

The Chinese francolin nests on the ground concealed in grasses and bushes. The clutch size ranges from 3 to 6 eggs, and eggs are creamy white without speckles.

Chinese Grouse

Bonasa sewerzowi / Ban Wei Zhen Ji
Severtzov's Grouse

Adult male

IUCN: NT
CITES: NL
CPS: I
CRL: NT (evaluated as *Tetrastes sewerzowi*)

BL: 33–40 cm
TL: 9–16 cm
WT: 200–400 g

Distribution

The Chinese grouse is endemic to China and restricted to forested mountains in central and southwest China. Within the range of this guide, Chinese grouse reside in southeast Qinghai and southern Gansu Provinces, northwest Yunnan Province, western Sichuan Province, and eastern Tibet AR.

Appearance

The Chinese grouse is a small pheasant, and both sexes are similar in appearance. The male has a black throat, bordered with white, a conspicuous white streak behind the eye, and a conspicuous patch of red bare skin above the eye. It has a short, bushy crest on its head. The body plumage is chestnut to brown, with upperparts barred with black and underparts boldly barred with black and white. The female has a chestnut throat with thin white streaks.

Habitat and Ecology

Chinese grouse inhabit conifer-dominated forests, alpine shrub, and grass-

Adult female.

Nest

land near the tree line between 2,000 and 4,200 m elevation. They feed on the ground and in trees and shrubs and especially favor birch stands and willow thickets along rivers. Buds, twigs, and shoots of willows and birches are their primary food; they also eat seeds, berries, and insects. The species is monogamous; the birds mate from late March to May, and the females lay eggs from May to June. They are often encountered as solitary birds or in pairs but can form foraging flocks up to 20 individuals during the winter.

Nest and Eggs

Chinese grouse build nests on bare ground at the base of trees. The clutch size ranges from 5 to 8 (average of 6), and the eggs are pale brown with dark chestnut speckles. The female is responsible for incubation, which takes about 27–29 days, and the male stays close to the nest site.

Green Peafowl
Pavo muticus / Lv Kong Que

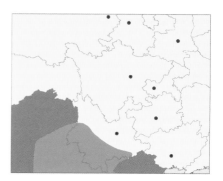

Adult male

IUCN: EN
CITES: II
CPS: I
CRL: CR

BL: 100–300 cm
TL: 100–160 cm
 (with train)
WT: 1–5 kg

Distribution

The green peafowl was once common-ly distributed in Southeast Asia, but its population has undergone a serious re-cent decline and has been much frag-mented. Within the range of this guide, green peafowl reside in southern Yun-nan Province.

Appearance

The green peafowl is a large, beautiful bird and the only wild peafowl species in China. Both sexes are similar in ap-pearance. The green peafowl has a long neck and upright crest atop its head and a yellow crescent beside the ear; the neck, breast, and upper back are ir-idescent green; the wings are blackish brown, and wing coverts are blue. The primaries are buff and visible during flight. The male has extremely long up-per tail coverts with numerous ocel-li that cover its actual tail underneath. When the males molt their train outside of breeding season, both sexes look quite similar. Compared with the male, the female has a slightly wider and shorter head crest and a reddish-brown loral stripe, which is black in the male.

Habitat and Ecology

Green peafowl are primarily found in tropical and subtropical forests in low mountains and hills or along valleys be-low an elevation of 2,500 m. They pre-fer open forest but also occur in bam-boo, shrubs, and farmland. They feed on seeds, fruits, insects, frogs, and rep-tiles. The species appears to be polyg-amous and breeds from February to June. The males possess a spectacu-lar display with the train fully fanned to reveal ocelli. The birds form family groups after the breeding season that usually consist of one male, several fe-males, and juveniles. They roost in trees at night, often near water. They are tim-id and hard to observe in the field; in most cases they are detected by calls from the roosting site. They have been widely raised by zoos and farms in China, where hybrids with Indian peafowl may occur when the two species are kept together.

Nest and Eggs

The green peafowl constructs its nest on the ground hidden by thick bushes. The nest is a shallow pit filled with twigs, leaves, grass, and feathers. Its clutch size ranges from 3 to 6 eggs, and the eggs (about 75 × 55 mm) are pale cream to buff without speckles. Only the female incubates the eggs, for 27–30 days.

Subadult male

Tibetan Snowcock
Tetraogallus tibetanus   Zang Xue Ji

Adult

IUCN: LC
CITES: NL
CPS: II
CRL: NT

BL: 46–64 cm
TL: 15–18 cm
WT: 1.1–1.8 kg

Distribution

The Tibetan snowcock is mainly distributed on the Qinghai-Tibetan Plateau, extending eastward to the Qilian Mountains and westward through the Pamirs in Tajikistan, India, Nepal, and Bhutan. This species is distributed at the highest elevations among the five species of *Tetraogallus*. Within the range of this

Female

Nest

guide, it resides in Qinghai Province and western Sichuan Province, as well as the Tibet AR.

Appearance

The Tibetan snowcock is a large, gray, partridge-like bird. Its crown and neck are gray, with a white throat and white crescent patch below the eye. The orbital skin and beak are red. The upperparts are sandy gray with buff streaks on the scapular and wing coverts, and the underparts are white with long black stripes on the flanks and belly. The bird's secondary feathers are broadly tipped with white and visible in flight; the rump and tail are rufous brown. The female has a similar appearance to the male, except it is smaller and duller and lacks the tarsal spur of the male.

Habitat and Ecology

Tibetan snowcocks occur on rocky slopes and in alpine meadows above the tree line between 3,000 and 6,000 m. They descend to scrubs and stunted cypress during severe winters but do not enter forest or dense shrubs. They feed primarily on leaves, flowers, roots, seeds, berries, and occasionally insects. They are seen following ungulates and forage on open ground trampled by those large animals when heavy snow is present. The birds are bulky and seldom fly. They quickly run uphill when they perceive danger, ready to glide downslope if necessary. They are good at gliding from one mountain ridge to another. The species appears to be monogamous and breeds from April to July. Foraging groups outside the breeding season can be up to 50 individuals. They roost among rocks and stones.

Nest and Eggs

Tibetan snowcocks construct their nest on the ground, often concealed by rocks and bushes. The nest, 28–47 cm wide and 6–12 cm deep, is a bare ground scrape filled with dry grass, moss, and feathers. The clutch size ranges from 4 to 7 eggs. The eggs are pale grayish blue or stony olive spotted with cinnamon brown. The female incubates the eggs alone, and eggs hatch after 27–28 days.

COMMON DOMESTIC ANIMALS AND OTHER WILDLIFE

COMMON DOMESTIC ANIMALS

Some tracks and sign encountered throughout the range of this guide belong to domestic species or to species smaller than those covered earlier in this guide. We assume that readers know the appearance of each domestic species but provide common sign and tracks in order to differentiate them from those of wildlife. For select small species, we provide sign that are common in forested regions of this guide.

Domestic dog (Tibetan mastiff) scats after foraging yak carcass

Dog

Canis familiaris // Gou

and does not contain as much hair or bone because of the dog's diet.

Tracks

The track size varies on the basis of breed. However, the shape of the track is generally oval and consists of four parallel rounded or egg-shaped toe prints and a triangle- or oval-shaped footpad. It is difficult to distinguish a large dog track from a wolf track on the basis of a single track. In contrast to felid tracks, dog tracks usually have claw marks in front of toe prints. In addition, the leading edge of the footpad print may show only one lobe, rather than the flat or multi-lobed felid tracks.

Scat

Dogs fed commercial dog food produce consistent brownish tubular scats with little or no obvious animal content. The scats vary greatly in size, color, and content because of the different breeds, diets, and life histories (e.g., semiferal dog). Domestic dog scat can usually be distinguished from that of other canine species as it often lacks a tapered end

Other Sign

Dogs produce a wide range of familiar vocalizations, including bark, whine, whimper, howl, huff, growl, and yelp.

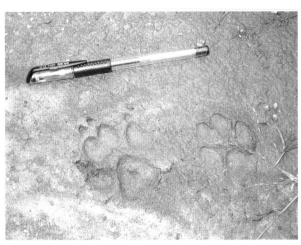
Domestic dog tracks

Domestic Cattle
Bos taurus // Huang Niu

Water Buffalo
Bubalus bubalis // Shui Niu

Yak
Bos grunniens // Mao Niu

Domestic cattle fecal pellets

Domestic cattle scat pile

Scat

Domestic cattle, water buffalo, and yaks produce large (>20 cm across) brownish scats. Young may have smaller scat, but it will be found in conjunction with adult scat. Large piles of scats are frequently found at bedding and feeding sites and at mineral licks. Their scats normally include clumped large fecal pellets during dry and cold seasons or a single pile that is flat and lacks structure during wet and hot seasons. The only similar scats are from takins and wild yaks, but those are more pellet shaped in a tubular form. The scats of yak calves are separate fecal pellets similar to those of wild ungulates but are usually found close to adult scats. Although cattle, water buffalo, and yaks produce similar scats and tracks, their difference in distribution altitude can be used for identification. Water buffalo are usually found below 1,000 m, domestic cattle are distributed from 800 to 3,200 m, and domestic yaks are found above 3,000 m.

Tracks

The animals often form large herds, leaving multiple tracks, feces, and feeding signs along the trail. Their footprints have a double-hoof imprint; each hoof has a narrower front tip and a round-shaped back edge. The tracks of domes-

Domestic water buffalo tracks

tic cattle, water buffalo, and yaks may look similar to those of takins or wild yaks but are easy to distinguish from those of smaller ungulates. In contrast to the heart-shaped tracks left by boars or deer, cattle tracks are usually larger and rounder, with no dewclaw marks. Track dimensions are approximately 10–15 cm long by 9–14 cm wide for adults.

Other Sign

A herd of these species in a forest or shrubs will leave obvious signs when they feed on bark, twigs, and leaves as high as they can reach. These species graze on long grass by wrapping their long tongues around the grass and pulling, producing a ragged appearance to the remaining grass. Other ungulate species will leave a more even leaf edge in their browsing and grazing.

Goat

Capra aegagrus hircus   Shan Yang

Scat

Scats are piles of small pellets (1–2 cm long by 0.5–1 cm wide). Each pellet is relatively uniform in size and shape. Pellets are round/oval shaped, smooth, and dark brownish/black. When they are fresh, they are glossy and often separate. As the pellets age, they become dull and deteriorate along their exposed edges.

Tracks

Prints show two symmetrical hooves with slightly incurved tips. The prints of goats are more kidney shaped (be-cause of the curve on the inside of the toe) with more pointed tips than those of pigs or deer. Prints are approximately 4.5–6 cm long by 2–3 cm wide for adults. Goat tracks usually do not show dewclaw marks unless in deep snow or mud, which make them distinguishable from tracks of small deer.

Other Sign

Goats are browsers and usually leave cleanly cut leaves and shoots as feeding sign. They are able to stand on their hind legs and reach vegetation that is more then 2 m high.

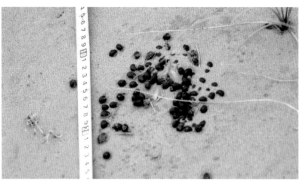

Domestic goat scat

Horse
Equus caballus / Ma

Domestic horse feces

shaped gap in the back. The prints are approximately 12–16 cm long by 8–14 cm wide for adults.

Other Sign

Horses can sleep both standing and lying down. When lying down, a group of horses will leave shallow indentations on the ground. Horses have both upper and lower incisors, which enable them to graze by clipping grass close to the ground with a clean edge.

Scat

Scats are usually large greenish pellets (5–8 cm long by 3–5 cm wide) that either clumped into a large pile or separately spread on the ground. Their scats are composed of partially digested vegetable material.

Tracks

In contrast to the double-hoof imprints left by ungulates, horses (as well as donkeys and mules) leave single-hoof footprints. Horses have symmetrical oval-shaped prints with a wedge-

Domestic horse tracks

OTHER WILDLIFE

Besides the large mammals and pheasants within this region, other wild-life species also leave sign in the wild. Here we present several examples of major groups, especially for the commonly observed species and those that leave specific signs that may cause confusion.

Bat

Bats leave droppings (called guano) under their resting sites in caves, rock crevices, tree cracks, and human building roofs. Fresh droppings are brown to black fecal pellets (0.2 × 0.7-1.5 cm) in a rice-like shape. Individual bat pellets are indistinguishable from those of rodents (e.g., rats), but bat droppings are typically found in large numbers at varied

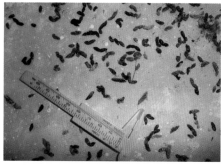

Bat droppings

ages right under their roosting sites, and usually, the roosting bats themselves can be spotted.

Pika

Pikas are a common small mammal (weight range of 50-350 g) living in both forest and alpine habitats (2,200-4,500 m) throughout southwest China. They are close relatives to hares and rabbits in the order Lagomorpha. Pikas live in burrows or in rock crevices. They have established latrines within their territory containing small, round pellets (0.2-0.3 cm in diameter).

A moupin pika (*Ochotona thibetana*) in its typical alpine open habitat

Burrow and latrine site of moupin pika in alpine meadow

Squirrels

Squirrels feed on mast seeds and fruits and leave various feeding signs such as broken mast shells, especially during the fall. Flying squirrels can feed on buds, twigs, and young bark and then leave large amounts of feeding signs such as small broken branches under the tree or shrub. These branches may be mistakenly recognized as feeding signs of primates (mostly snub-nosed monkeys), but they are characterized by white, debarked wood with dense, tiny teeth marks. The unique round fecal pellets, varying from yellow to brown or black, are frequently found nearby.

The most commonly encountered flying squirrel species in southwestern China, the gray-headed flying squirrel (*Petaurista caniceps*)

Large collection of walnut shells left by squirrels under a rock

Broken branches with missing bark and typical round fecal pellets left by the flying squirrel *Trogopterus xanthipes*

Walnut shells left by Pere David's rock squirrel (*Sciurotamias davidianus*)

Bamboo Rat

Bamboo rats are a group of burrow-dwelling rodents whose diet is composed almost entirely of bamboo. They are rarely observed, but their burrows (typically elevated piles of disturbed soil) can be found in forests with dense bamboo understory, usually with feeding signs on bamboo shoots and stems nearby. Compared with the feeding sign left by the giant panda, bamboo rats typically cut the bamboo stems near the base, and the cut edge is smooth.

Feeding site of Chinese bamboo rat (*Rhizomys sinensis*), the most common bamboo rat species within the region. Note the bamboo stem to the left was cut, and the soil to its right was disturbed as the bamboo rat come out from underground burrow.

Lesser bamboo rat (*Cannomys badius*) at its burrow entrance

Owls

Owls and other birds of prey (e.g., hawks, eagles, shrikes) produce bird pellets as they regurgitate the undigested animal remains. Such pellets are typically round or a single segment composed of rodent hair, bone, skin, or bird feathers. They are normally found under the roosting tree or shrub and can be distinguished from carnivore scat as such pellets are always accompanied by the white droppings of birds.

Pellets produced by the Eurasian eagle owl (*Bubo bubo*)

Bee-Eaters and Martins

Some birds (e.g., bee-eaters, *Merops* spp., and sand martins, *Riparia riparia*) construct nests on vertical sandy banks along rivers and cliffs. Such nests appear similar to mammal burrows.

Hole nests of blue-tailed bee-eater (*Merops philippinus*) on a vertical river bank

Woodpeckers

The foraging holes of woodpeckers left on tree trunks are commonly seen in forests. The arrangement of holes can be species specific; those of the three-toed woodpecker, *Picoides tridactylus,* are easily recognized. Woodpeckers make shallow holes in the bark to feed on the tree sap in coniferous forests; the result is numerous evenly spaced holes in parallel rings on the trunk.

Holes left by a three-toed woodpecker on a large conifer tree

IMAGE CREDITS

Page **Image**

10 A mosaic of agriculture and forest. Pingwu County, Sichuan Province. Photo by William McShea.

10 Rice paddies. Medog County, Tibet AR. Photo by Wang Fang.

11 A monoculture of young planted fir (*Abies*) trees. Zhaotong City, Yunnan Province. Photo by Wang Fang.

11 Rubber plantation. Jinghong City, Yunnan Province. Photo by Li Sheng.

11 High elevation grassland. Sanjiangyuan NR, Qinghai Province. Photo by Wu Lan.

12 Alpine valley with schree. Bomi County, Tibet AR, 2009. Photo by Wang Fang.

12 High elevation grass/shrubland valley. Sichuan Province. Photo by Wang Fang.

13 Semidesert grassland common on northern plateau area. Qinghaihu NR, Qinghai Province, 2009. Photo by Wang Fang.

13 Alpine meadows embedded within forested mountains. Batang County, Sichuan Province. Photo by Dong Lei.

14 Coniferous forest with evergreen understory. Wanglang NR, Sichuan Province. Photo by Li Sheng.

14 Mixed deciduous and evergreen forest usually at sites with abundant rainfall or fog. Medog County, Tibet AR. Photo by Wang Fang.

15 Mixed deciduous/coniferous forest with a relatively high diversity of tree species, usually abundant water in narrow streams. Laohegou NR, Sichuan Province. Photo by Wang Fang.

15 Alpine oak (*Quercus*) forest at high elevation. Gexigou NR, Sichuan Province. Photo by Li Sheng.

15 Coniferous forest. Wanglang NR, Sichuan Province. Photo by Li Sheng.

15 Female wild boar with offspring in young forest opening with bamboo understory. Changqing NR, Shaanxi Province. Photo by Sheng Li.

15 Tropical forest. Medog County, Tibet AR, 2009. Photo by Wang Fang.

18 Illustrations, animal tracks and features. Created by Jiazi Liu.

24 Adult male black-crested gibbon. Wuliangshan NR, Yunnan Province. 2011. Photo by Zhang Xingwei.

24 Adult female black-crested gibbon; note the black streak on its head. Wuliangshan NR, Yunnan Province. 2011. Photo by Zhang Xingwei.

25 Adult male eastern hoolock gibbon. Gaoligongshan NR, Yunnan Province. 2010. Photo by Xi Zhinong.

26 Adult female (right) and subadult (left) eastern hoolock gibbon. Gaoligongshan NR, Yunnan Province. 2010. Photo by Li Binbin.

26 Hoolock gibbon habitat; note the contiguous canopy. Gaoligongshan NR, Yunnan Province. 2010. Photo by Wang Fang.

26 Hoolock gibbon scat. Gaoligongshan NR, Yunnan Province. 2014. Photo by Li Jiahong.

26 Hoolock gibbon scat. Gaoligongshan NR, Yunnan Province. 2014. Photo by Li Jiahong.

26 Hoolock gibbon feeding sign. Gaoligongshan NR, Yunnan Province. 2014. Photo by Li Jiahong.

27 Adult male northern white-cheeked gibbon (captive individual). Xishuangbanna, Yunnan Province. 2007. Photo by Wang Fang.

29 Adult female François' langur and two young; note the bright orange pelage of the young. Mayanghe NR, Guizhou Province. 2015. Photo by Xu Jianmin.

Page	Image
30	Adult Nepal gray langur. Mêdog County, Tibet AR. 2013. Photo by Zuo Lingren.
31	Adult female Shortridge's langurs. Gongshan County, Yunnan Province. 2013. Photo by Peng Jiansheng/IBE.
32	Two females and a newborn white-headed langur. Chongzuo County, Guangxi AR. 2012. Photo by Wang Fang.
32	White-headed langur. The rock face shows the common brown urine stains visible when the species is present. Chongzuo County, Guangxi AR. 2012. Photo by Wang Fang.
33	White-headed langur scat. Chongzuo County, Guangxi AR. 2015. Photo by Li Sheng.
33	Footprint of white-headed langur (hind foot); note the long and narrow shape. Chongzuo County, Guangxi AR. 2015. Photo by Liang Zuhong.
33	Sleeping site of white-headed langur; note the brownish urine stains on the limestone hills in Chongzuo County, Guangxi AR. 2012. Photo by Wang Fang.
34	White-headed langur habitat; note the fragmented forest patches on limestone cliffs isolated by agricultural lands. Chongzuo County, Guangxi AR. 2011. Photo by Wang Fang.
34	Track of white-headed langur (forefoot). Chongzuo County, Guangxi AR. 2015. Photo by Li Sheng.
35	Phayre's leaf monkey. Gaoligongshan NR, Yunnan Province. 2012. Photo by Li Jiahong.
35	Phayre's leaf monkey and its habitat. Gaoligongshan NR, Yunnan Province. 2012. Photo by Li Jiahong.
37	Assam macaque; note the brownish facial skin. Gaoligongshan NR, Yunnan Province. 2013. Photo by Li Jiahong.
37	Assam macaque scat composed of seeds. Gaoligongshan NR, Yunnan Province. 2014. Photo by Lin Rutao.
38	Assam macaque habitat. Gaoligong NR, Yunnan Province. 2011. Photo by Wang Fang.
38	Adult male northern pig-tailed macaque. Khao Yai National Park, Thailand. 2011. Photo by Zhao Chao.
39	Female rhesus macaque and subadults; note the reddish thinly haired facial skin. Gongbujiangda County, Tibet AR. 2008. Photo by Wang Fang.
39	Juvenile rhesus macaque. Yangxian County, Shaanxi Province. 2010. Photo by Hu Wanxin.
40	Rhesus macaque's tracks; note the thumb pointing at a right angle. Changqing NR, Shaanxi Province. 2014. Photo by Xiang Dingqian.
40	Stump-tailed macaque. Gaoligongshan NR, Yunnan Province. 2015. Photo by KFBG-KCC.
41	Tibetan macaque. Tangjiahe NR, Sichuan Province. 2009. Photo by Li Sheng.
42	Tibetan macaque scat; note the fibrous plant material. Tangjiahe NR, Sichuan Province. 2014. Photo by Li Sheng.
42	Wild kiwi fruits partly eaten by Tibetan macaque. Tangjiahe NR, Sichuan Province. 2015. Photo by Li Sheng.
42	White-cheeked macaque and its tropical forest habitat. Mêdog County, Tibet AR. 2014. Photo by Li Cheng.
43	White-cheeked macaque, with white, elongated whiskers on the face and dark facial skin. Mêdog County, Tibet AR. 2015. Photo by Li Cheng.
45	Bengal slow loris. Gaoligong NR, Yunnan Province. 2013. Photo by Li Jiahong.
46	Pygmy slow loris. Pingbian County, Yunnan Province. 2015. Photo by Zhang Qi.
46	Pygmy slow loris scat. Pingbian County, Yunnan Province. 2015. Photo by Zhang Qi.

Page	Image
48	Adult male black snub-nosed monkey. Baimaxueshan NR, Yunnan Province. 2010. Photo by Xi Zhinong.
48	Black snub-nosed monkey scat; note the discrete pellets. Laojunshan NR, Yunnan Province. 2014. Photo by Jin Tong.
49	Black snub-nosed monkey habitat, primary forest above 4,000 m elevation in Baimaxueshan NR, Yunnan Province. 2011. Photo by Xi Zhinong.
49	Adult male golden snub-nosed monkey. Huangbaiyuan NR, Shaanxi Province. 2011. Photo by Wang Fang.
50	Adult (front) and subadult (in back, without the lappets at the corner of mouth) male golden snub-nosed monkeys. Laohegou NR, Sichuan Province. 2016. Photo by Li Sheng.
50	Golden snub-nosed monkey scat; note the connected pellets. Laohegou NR, Sichuan Province. 2005. Photo by Li Sheng.
50	Golden snub-nosed monkey track. Huangguanshan NR, Shaanxi Province. 2010. Photo by Wang Fang.
50	Branch broken by golden snub-nosed monkey. Tangjiahe NR, Sichuan Province. 2014. Photo by Li Sheng.
50	Golden snub-nosed monkey feeding site; note the barkless branches in the trees. Tangjiahe NR, Sichuan Province. 2014. Photo by Li Sheng.
51	Adult female (top) and male (bottom) gray snub-nosed monkeys; note the brighter tones on the male. Fanjingshan NR, Guizhou Province. 2006. Photo by Wang Fang.
51	Subadult gray snub-nosed monkey foraging on flowers. Fanjingshan NR, Guizhou Province. 2006. Photo by Wang Fang.
52	Adult male Myanmar snub-nosed monkey. Gaoligongshan NR, Yunnan Province. 2013. Photo by Zuo Lingren.
52	Adult male (left) and female (right) Myanmar snub-nosed monkeys. Gaoligongshan NR, Yunnan Province. 2013. Photo by Zuo Lingren.
53	Adult dhole in broadleaf forest. Mêdog County, Tibet AR. 2014. Photo by Li Cheng/IBE.
54	Scat of dhole. Mêdog County, Tibet AR. 2014. Photo by Li Cheng.
54	Track of dhole. Salakpra Wildlife Reserve, Thailand. 2015. Photo by N. Songsasen.
55	Adult raccoon dog. Shanghai. 2015. Photo by Sun Xiaodong.
55	Raccoon dog scat. Chernobyl Exclusion Zone, Belarus. 2015. Photo by Cara Love.
56	Adult red fox in its summer coat in the alpine grassland habitat. Ali Prefecture, Tibet AR. 2013. Photo by Guo Liang/IBE.
57	Adult red fox in its winter coat in the alpine rocky habitat at 4,300 m elevation. Wolong NR, Sichuan Province. 2009. Photo by Li Sheng.
57	Fresh red fox scat. Note the absence of needle-shaped ends. Shenguozhuang NR, Sichuan Province. 2009. Photo by Li Sheng.
58	Red fox burrow. Yajiang County, Sichuan Province. 2016. Photo by Li Sheng.
58	Old Himalayan marmot burrow that was temporarily used by a red fox as evidenced by the hairs and scat (the dry white scat inside the burrow) left at the burrow entrance. Yajiang County, Sichuan Province. 2016. Photo by Li Sheng.
58	Adult Tibetan fox in its summer coat. Sanjiangyuan NR, Qinghai Province. 2012. Photo by Zhang Yong.
59	Tracks of Tibetan fox on snow. Sanjiangyuan NR, Qinghai Province. 2011. Photo by Bu Hongliang.
59	Scat of Tibetan fox. Xinjiang AR. 2006. Photo by Liu Yanlin.
60	Adult wolf in its winter coat. Sanjiangyuan NR, Qinghai Province. 2011. Photo by Zhou Jie.

Page	Image
60	Adult wolf in its summer coat. Sanjiangyuan NR, Qinghai Province. 2013. Photo by Dong Lei/IBE.
61	Wolf track left on sandy ground. Note the clear imprints of claws. Sanjiangyuan NR, Qinghai Province. 2011. Photo by Bu Hongliang.
61	Wolf track left on snow. Sanjiangyuan NR, Qinghai Province. 2012. Photo by Zhou Jie.
61	Fresh wolf scat. Yajiang County, Sichuan Province. 2016. Photo by Li Sheng.
61	Old wolf scat. Yajiang County, Sichuan Province. 2016. Photo by Li Sheng.
63	Side view of adult Asiatic black bear. Note the stocky body, round ears, small eyes, and the short tail. Laohegou NR, Sichuan Province. 2015. Photo by Li Sheng.
64	Juvenile Asiatic black bear. Note the V-shaped white patch on the chest. Laohegou NR, Sichuan Province. 2012. Photo by Li Sheng.
64	Scat of Asiatic black bear after feeding on berries during fall. The pie-shaped scat is loose and moist, containing numerous berry seeds. Tangjiahe NR, Sichuan Province. 2006. Photo by Li Sheng.
65	Hind foot track of Asiatic black bear left on snow. Jiudingshan NR, Sichuan Province. 2012. Photo by Dong Lei.
65	Forefoot track of Asiatic black bear left on soft soil. Wanglang NR, Sichuan Province. 2015. Photo by Luo Chunping.
65	Fresh scat of Asiatic black bear with a diet of both animal and plant items. Xiaohegou NR, Sichuan Province. 2008. Photo by Li Sheng.
65	Fresh scat of Asiatic black bear with a diet of primarily acorn mast during late fall. Changqing NR, Shaanxi Province. 2009. Photo by Xiang Dingqian.
66	Claw marks of Asiatic black bear on alpine oak tree with coarse bark. The marks appear as pointed scratches. Xiaohegou NR, Sichuan Province. 2008. Photo by Li Sheng.
66	Feeding sign of Asiatic black bear searching for ant nests under logs. Heishui County, Sichuan Province. 2006. Photo by Shao Liangkun.
66	Ant nest and eggs under a rock disturbed by Asiatic black bear. Yongzhongling NR, Sichuan Province. 2006. Photo by Li Sheng.
67	Feeding platform left by Asiatic black bear on a deciduous broadleaf tree. Yajiang County, Sichuan Province. 2016. Photo by Li Sheng.
67	Claw marks of Asiatic black bear on tree trunk with soft and smooth bark. The marks appear as long, parallel scratches. Xiaohegou NR, Sichuan Province. 2008. Photo by Li Sheng.
67	Feeding sign of Asiatic black bear on a coniferous tree after it removed the bark to scrape the tree sap and soft wood under the bark. Luhuo County, Sichuan Province. 2006. Photo by Li Sheng.
68	Adult brown bear scavenging on yak carcass in the alpine grassland of the Qinghai-Tibet Plateau. Note the large whitish patch from shoulder to chest and down to the belly. Kekexili NR, Qinghai Province. 2015. Photo by Feng Jiang/ Nature Image China.
69	Brown bear scat as a flat pie-shaped mass found in the conifer forest near the tree line. The scat contained undigested skin, bones, and hairs of Himalayan marmots. Shenxianshan NR, Sichuan Province. 2006. Photo by Li Sheng.
69	Brown bear scat as segments linked by animal hairs found in the alpine grassland. The scat contained mainly ungulate hairs. Sanjiangyuan NR, Qinghai Province. 2011. Photo by Wu Lan.
70	Forefoot track of brown bear left on sandy ground. Sanjiangyuan NR, Qinghai Province. 2011. Photo by Wu Lan.
70	Hind foot track of brown bear left on ground with loose sand and small rocks. Note the track has no clear edge and may look like a human footprint. Sanjiangyuan NR, Qinghai Province. 2011. Photo by Wu Lan.

Page	Image
70	Bedding site of brown bear under rock cliff. Sanjiangyuan NR, Qinghai Province. 2011. Photo by Wu Lan.
70	Captive adult brown bear at Xining Zoo, Qinghai Province. 2010. Photo by Xiang Dingqian.
71	Brown bear tracks left on ground with snow. Note the obvious difference between the forefeet (first print on the left) and the hind feet (second print on the left). Sanjiangyuan NR, Qinghai Province. 2012. Photo by Wu Lan.
71	Adult giant panda with regular black-and-white coat color sniffing at the scent mark on a tree trunk along a ridge trail. Changqing NR, Shaanxi Province. 2008. Photo by PKU/eMammal.
72	Giant panda rubbing against a conifer tree during the mating season to deposit secretions from the anal glands located under its tail. Wanglang NR, Sichuan Province. 2011. Photo by PKU/eMammal.
72	Adult giant panda with rare brown-and-white coat color in Qinling Mountains. Niuweihe NR, Shaanxi Province. 2013. Photo by PKU/eMammal.
73	Fresh scat of giant panda containing primarily parts of bamboo stems. Wolong NR, Sichuan Province. 2006. Photo by Li Sheng.
73	Fresh scat of giant panda containing primarily parts of bamboo stems. Wolong NR, Sichuan Province. 2006. Photo by Li Sheng.
73	Large piles of giant panda scats left at a feeding site. Laohegou NR, Sichuan Province. 2005. Photo by Li Sheng.
73	Rare giant panda scat containing animal remains. The right one was broken and contained bamboo leaves, as well as hair, bone, and nails of a golden snub-nosed monkey. The left one was unbroken; note the distinct white coating. Kuanba Timber Area, Pingwu County, Sichuan Province. 2014. Photo by Li Sheng.
74	Giant panda track in soft soil. Wolong NR, Sichuan Province. 2006. Photo by Li Sheng.
74	Giant panda track left on frozen snowy ground (left forefoot, right hind foot). Wanglang NR, Sichuan Province. 2004. Photo by Li Sheng.
74	Fresh feeding sign and scats left by giant panda after feeding on new bamboo shoots. Wolong NR, Sichuan Province. 2006. Photo by Li Sheng.
74	Small pile of bamboo shoot and stem remains left by giant panda after feeding. Wolong NR, Sichuan Province. 2006. Photo by Li Sheng.
75	Scent mark left by a giant panda on a dead conifer during mating season. Note there are bite marks and hairs in the middle of the mark. Changqing NR, Shaanxi Province. 2008. Photo by Li Sheng.
75	Birthing den of giant panda in a rock crevice. Foping NR, Shaanxi Province. 2010. Photo by Li Yunxi.
75	Birthing and/or nursing den within a hollow fir tree recently used by giant pandas. Note the large pile of scat outside of the den. Erlangshan Forestry Area, Sichuan Province. 2015. Photo by Li Yunxi.
76	Adult red panda. Anzihe NR, Sichuan Province. 2010. Photo by Fu Qiang.
77	Red panda track from captive animal. Front Royal, Virginia, USA. 2016. Photo by Jessica Kordell/SCBI.
77	Fresh red panda scats left in conifer forest with dense bamboo understory. Wolong NR, Sichuan Province. 2007. Photo by Li Sheng.
78	Old scat of Siberian weasel in its typical twist rope-like form. Changqing NR, Shaanxi Province. 2008. Photo by Li Sheng.
78	Fresh scat of Siberian weasel. Tangjiahe NR, Sichuan Province. 2009. Photo by Li Sheng.
79	Family group of Asian badgers. Hongyuan County, Sichuan Province. 2006. Photo by Yin Yufeng.

Page	Image
80	Dry Asian badger scat in its typical form. Note the ball of undigested beetle shells in the segment to the right. Yajiang County, Sichuan Province. 2016. Photo by Li Sheng.
80	Pile of scats left by Asian badger at its latrine site. Note the pile contains scats at varied ages. Qinghaihu NR, Qinghai Province. 2009. Photo by Liu Jiazi.
81	Track of Asian badger left on wet sandy ground. Note the deep imprints of claws. Qinghaihu NR, Qinghai Province. 2009. Photo by Liu Jiazi.
81	Burrow entrance of Asian badgers on level ground in semiarid steppe grassland habitat. The researcher is pointing to a latrine site. Qinghaihu NR, Qinghai Province. 2009. Photo by Liu Jiazi.
81	Adult Chinese ferret badger. Changqing NR, Shaanxi Province. 2012. Photo by Hu Wanxin.
83	Adult northern hog badger. Wanglang NR, Sichuan Province. 2012. Photo by Li Sheng.
83	Adult female northern hog badger and her offspring. Wanglang NR, Sichuan Province. 2006. Photo by Wang Fang.
84	Fresh scat of northern hog badger in a typical curled shape. Tangjiahe NR, Sichuan Province. 2016. Photo by Li Sheng.
84	Old dry scats of northern hog badger in a typical curled shape. Shenxianshan NR, Sichuan Province. 2006. Photo by Li Sheng.
84	Digging pit left by northern hog badger during fall in deciduous broadleaf forest. Tangjiahe NR, Sichuan Province. 2015. Photo by Li Sheng.
85	Adult beech marten. Wolong NR, Sichuan Province. 2009. Photo by Li Sheng.
86	Adult yellow-throated marten. Laohegou NR, Sichuan Province. 2012. Photo by PKU Wildlife Lab.
86	Pair of adult yellow-throated martens. Jiudingshan NR, Sichuan Province. 2004. Photo by PKU/eMammal.
87	Fresh scat of yellow-throated marten found in conifer forest at 3,000 m elevation, containing mostly hairs of small insectivores and rodents. Wanglang NR, Sichuan Province. 2015. Photo by Li Sheng.
87	Old scat of yellow-throated marten left on elevated rock surface, containing primarily rodent bone fragments and hairs. Laohegou NR, Sichuan Province. 2013. Photo by Li Sheng.
87	Scat from yellow-throated marten after feeding on a bee nest. The scat is light brownish compared to the regular black color. Gexigou NR, Sichuan Province. 2006. Photo by Li Sheng.
88	Captive adult Asian small-clawed otter at the National Zoological Park, Washington, D.C., USA. 2005. Photo by Li Sheng.
89	Adult Eurasian otter. Tangjiahe NR, Sichuan Province. 2014. Photo by Ma Wenhu.
90	Fresh footprints of Eurasian otter on muddy ground near a stream. Jiuzhaigou NR, Sichuan Province. 2015. Photo by Zhu Lei.
90	Scats of Eurasian otter found on top of rock. Note this pile contains multiple scats of varied ages and undigested fish bones. Sanjiangyuan NR, Qinghai Province. 2013. Photo by Tashi Sanger.
90	Latrine of Eurasian otter on rock along riverside with scats of varied ages. Pingwu County, Sichuan Province. 2015. Photo by Shao Liangkun.
90	Track of Eurasian otter left on snow. Note that the imprints of the palm pad have no clear edges because of the presence of a web between the toes. Sanjiangyuan NR, Qinghai Province. 2014. Photo by He Bing.
91	Adult mountain weasel in its summer coat. Lhasa, Tibet AR. 2006. Photo by Dong Lei.

Page	Image
91	Typical alpine rocky habitat of mountain weasel and its paler winter coat. Note the brushy tail during winter. Wolong NR, Sichuan Province. 2010. Photo by Li Sheng.
92	Adult Siberian weasel in its winter coat. Jiudingshan NR, Sichuan Province. 2012. Photo by Dong Lei.
93	Pair of adult steppe polecats during summer. Note the side profile of arched body while jumping. Ruoergai NR, Sichuan Province. 2009. Photo by Dong Lei.
94	Front view of adult steppe polecats. Ruoergai NR, Sichuan Province. 2009. Photo by Dong Lei.
94	Steppe polecat with its prey (a pika) while exiting the burrow entrance. Note the much paler pelage on the head and body in its denser winter coat. Qinghai Province. 2008. Photo by Zuo Lingren.
95	Adult yellow-bellied weasel in its summer coat. Gutianshan NR, Zhejiang Province. 2014. Photo by Shen Xiaoli.
96	Track of masked palm civet left on muddy ground. Changqing NR, Shaanxi Province. 2013. Photo by Xiang Dingqian.
96	Remains of fruit that fell on the ground after being eaten by masked palm civets. Houhe NR, Hubei Province. 2005. Photo by Zhou Youbing.
98	Common palm civet foraging on trees during night. Yingjiang County, Yunnan Province. 2015. Photo by Li Fei.
99	Large Indian civet, front view. Bayuelin NR, Sichuan Province. 2015. Photo by Li Sheng.
99	Large Indian civet. Bayuelin NR, Sichuan Province. 2015. Photo by Li Sheng.
100	Masked palm civet. The fur appears light gray to whitish under the camera flash. Wanglang NR, Sichuan Province. 2011. Photo by Li Sheng.
101	Masked palm civet, side view. Wanglang NR, Sichuan Province. 2010. Photo by Li Sheng.
101	Den of masked palm civets. Houhe NR, Hubei Province. 2002. Photo by Zhou Youbing.
102	Scats of masked palm civet composed of both plant remains (as in the upper black segment) and animal remains (rodent hairs, as in the twisted gray segment to the left). The segments may have been defecated at different times. Houhe NR, Hubei Province. 2005. Photo by Zhou Youbing.
102	Fresh scats of masked palm civets as defecated near stream. Note the scat is in short segments and comprises undigested seeds, fruit skin, and shells. Houhe NR, Hubei Province. 2005. Photo by Zhou Youbing.
102	Typical scats of masked palm civets as defecated in shallow water of stream. Note the fresh scat is composed completely of undigested berry seeds. Houhe NR, Hubei Province. 2005. Photo by Zhou Youbing.
103	Adult small Indian civet. Hong Kong. 2014. Photo by Li Cheng
104	Adult spotted linsang. Gaoligongshan NR, Yunnan Province. 2016. Photo by KFBG-KCC.
105	Adult crab-eating mongoose. Note the pink nose, white stripe from the chin to the side of the neck, and the long toes. Chongzuo, Guangxi AR. 2012. Photo by Song Ye.
105	Adult crab-eating mongoose. Jinggangshan NR, Jiangxi Province. 2013. Photo by Song Dazhao.
106	Forefoot track of a crab-eating mongoose. Xishuangbanna NR, Yunnan Province. 2008. Photo by Sun Ge.
106	Hind foot track of a crab-eating mongoose. Xishuangbanna NR, Yunnan Province. 2008. Photo by Sun Ge.
107	Adult Javan mongoose in its winter coat. Hong Kong. 2016. Photo by Chen Kaiwen.

Page	Image
108	Adult clouded leopard. Nangunhe NR, Yunnan Province. 2007. Photo by Feng Limin.
109	Clouded leopard scat. Mêdog County, Tibet AR. 2013. Photo by Li Cheng.
110	Adult leopard in the high-elevation conifer forest during winter. Xinlong County, Sichuan Province. 2013. Photo by Gu Xiaodong.
111	Fresh claw mark left by leopard on tree during winter; a large piece of tree bark is missing. Heshun, Shanxi Province. 2008. Photo by Song Dazhao.
112	Fresh leopard scat, containing remains of musk deer (hairs and bone chips) and pheasants (feet and feathers). Note the short tail on the end of last segment (bottom). Luhuo County, Sichuan Province. 2006. Photo by Li Sheng.
112	Fresh leopard scat, containing hair of domestic yak. Note the tail feature is absent. Shenxianshan NR, Sichuan Province. 2006. Photo by Li Sheng.
113	Leopard track left on dry soil surface. Heshun, Shanxi Province. 2008. Photo by Song Dazhao.
113	Adult leopard scratching on a tree during early spring. Jiyuan NR, Henan Province. 2015. Photo by Xiao Zhishu.
113	Scratching pit left by leopard in pine forest during winter. Heshun, Shanxi Province. 2014. Photo by Song Dazhao.
114	Adult male snow leopard. Wolong NR, Sichuan Province. 2009. Photo by Li Sheng.
115	Old scats of snow leopard and its typical habitat of steep alpine rocky areas. Sanjiangyuan NR, Qinghai Province. 2014. Photo by Dong Lei/IBE.
116	Fresh track of snow leopard on muddy riverside during summer. Sanjiangyuan NR, Qinghai Province. 2014. Photo by Dong Lei/IBE.
116	Fresh scats of snow leopard during winter. Sanjiangyuan NR, Qinghai Province. 2010. Photo by Zhou Jie.
116	Scratching pit created by snow leopard. Sanjiangyuan NR, Qinghai Province. 2011. Photo by Wu Lan.
117	Fresh track of snow leopard on snow. Sanjiangyuan NR, Qinghai Province. 2010. Photo by Xiang Dingqian.
117	Fresh kill left by snow leopard. Sanjiangyuan NR, Qinghai Province. 2010. Photo by Xiang Dingqian.
117	Adult Indochinese tiger. The only and maybe the last photograph of this subspecies taken in the wild within China. Xishuangbanna NR, Yunnan Province. 2007. Photo by Feng Limin.
118	Track of adult tiger. Xishuangbanna NR, Yunnan Province. 2007. Photo by Feng Limin.
119	Leopard cat tracks left on muddy ground. The upper right is track made by the forefoot, and the lower left one is from the hind foot. Laohegou NR, Sichuan Province. 2013. Photo by Li Sheng.
119	Typical leopard cat scat with segments. Note the last segment has a long needle-shaped tail of plant fibers. The last segment is partly opened by the authors to show the rodent hairs and grasses. Luhuo County, Sichuan Province. 2006. Photo by Li Sheng.
120	Typical golden form of the Asiatic golden cat. Note the stripes on the forehead, faint spots on the legs and belly, and the distinctive curved black tail tip with bright white underside. Laohegou NR, Sichuan Province. 2013. Photo by Li Sheng.
121	Spotted form of Asiatic golden cat. Tangjiahe NR, Sichuan Province. 2008. Photo by Li Sheng.
121	Hind foot track of Asiatic golden cat. Laohegou NR, Sichuan Province. 2014. Photo by Li Sheng.
122	Asiatic golden cat scat (to the left of the GPS). Note that the lower end is composed of undigested grasses. Laohegou NR, Sichuan Province. 2013. Photo by Bu Hongliang.

Page	Image
122	Old dry segmented Asiatic golden cat scat under a rocky cliff. Laohegou NR, Sichuan Province. 2012. Photo by Li Sheng.
122	Front view of adult Chinese mountain cat during winter. Sanjiangyuan NR, Qinghai Province. 2010. Photo by Zhou Jie.
123	Flank view of adult Chinese mountain cat during summer. Hongyuan County, Sichuan Province. 2007. Photo by Yin Yufeng.
124	Adult leopard cat in its summer coat. Wanglang NR, Sichuan Province. 2010. Photo by Li Sheng.
124	Adult leopard cat in its winter coat. Wanglang NR, Sichuan Province. 2005. Photo by PKU/eMammal.
125	Latrine site for leopard cats with numerous scats at varied ages, from fresh (black one on the right with smooth surface) to old (central ones on the bottom that are dry, loose, and white). Wanglang NR, Sichuan Province. 2004. Photo by Li Sheng.
125	Fresh scat, track, and scratching pit on snow left by a leopard cat. Wanglang NR, Sichuan Province. 2004. Photo by Li Sheng.
125	Leopard cat scat. The longest segment contains grass fibers. Daofu County, Sichuan Province. 2006. Photo by Li Sheng.
125	Leopard cat track. Changqing NR, Shaanxi Province. 2008. Photo by Xiang Dingqian.
126	Adult lynx. Pulan County, Tibet AR. 2015. Photo by Guo Liang/IBE.
126	Scat left by lynx. Shenxianshan NR, Sichuan Province. 2016. Photo by Li Sheng.
127	Lynx track of front foot left on snow. Note the four toes are set apart and the shallow depression around the imprints of the pad and toes left by the long, thick footpad hairs. Daocheng County, Sichuan Province. 2009. Photo by Que Pinjia.
127	Old, white scats at a larine site used by lynx and other carnivores (wolves, leopard cats, etc.). Shenxianshan NR, Sichuan Province. 2016. Photo by Li Sheng.
127	Typical latrine shared by lynx (scat to the upper left of the coin) and other sympatric carnivores, including the leopard cat (black scats to the lower right of the coin) and wolf (old dry white scats to the lower left of the coin and to the right). Species identification was verified using fecal DNA analysis. Shenxianshan NR, Sichuan Province. 2016. Photo by Li Sheng.
128	Front view of adult marbled cat. Gaoligongshan NR, Yunnan Province. 2015. Photo by KFBG-KCC.
128	Side view of adult marbled cat. Note the tail is held horizontal with the back. Gaoligongshan NR, Yunnan Province. 2015. Photo by KFBG-KCC.
129	Adult Pallas's cat during summer. Zhiduo County, Qinghai Province. 2013. Photo by Zhang Yong.
130	Adult Pallas's cat near its den (the larger burrow entrance below the animal). Sanjiangyuan NR, Qinghai Province. 2012. Photo by Dong Lei/IBE.
131	Adult male Asian elephant. Xishuangbanna NR, Yunnan Province. 2012. Photo by Shen Qingzhong.
132	Herd of Asian elephants including mixed genders and ages gathering at river side to drink water and take a mud bath. Xishuangbanna NR, Yunnan Province. 2012. Photo by Shen Qingzhong.
133	Fresh scats of Asian elephants. Xishuangbanna NR, Yunnan Province. 2008. Photo by Sun Ge.
133	Old dry dung of Asian elephants. Xishuangbanna NR, Yunnan Province. 2008. Photo by Sun Ge.
133	Degraded Asian elephant scats after they fed on a rice field during the crop harvest season. Note the newly emerged seedlings germinated from undigested rice seeds in the dung. Xishuangbanna NR, Yunnan Province. 2008. Photo by Sun Ge.

Page	Image
133	Fresh track of Asian elephants left on muddy ground. Xishuangbanna NR, Yunnan Province. 2008. Photo by Sun Ge.
133	Track of the left hind leg of an Asian elephant left on wet soil. Note that there are several tracks left by a northern red muntjac within the elephant footprint. Xishuangbanna NR, Yunnan Province. 2008. Photo by Sun Ge.
134	Feeding sign left by Asian elephants after they fed on banana leaves and stems. Xishuangbanna NR, Yunnan Province. 2007. Photo by Sun Ge.
135	Mineral lick used by an Asian elephant herd. Note the tracks left on the slope. Xishuangbanna NR, Yunnan Province. 2008. Photo by Sun Ge.
135	Vertical bank of mineral lick recently visited by Asian elephants. Note smooth surface and the digging hole left by their tusks (pointed out by the field staff). Xishuangbanna NR, Yunnan Province. 2008. Photo by Sun Ge.
136	Fresh separated fecal pellets of forest musk deer. Changqing NR, Shaanxi Province. 2009. Photo by Xiang Dingqian.
136	Loosely clumped fresh fecal pellets of sambar found in moist low-elevation forest. Pellets are blunt and dimpled on one end. Anzihe NR, Sichuan Province. 2015. Photo by Fu Qiang.
137	Dry fecal pellets of sambar in high-elevation open habitat of alpine meadow. Yajiang County, Sichuan Province. 2016. Photo by Li Sheng.
137	Fresh fecal pellets of forest musk deer, with size comparison to tufted deer fecal pellets (older pellets on bottom of pile). Wanglang NR, Sichuan Province. 2013. Photo by Li Sheng.
137	Fresh fecal pellets of northern red muntjac. Xishuangbanna NR, Yunnan Province. 2008. Photo by Sun Ge.
137	Clumped fecal pellets of northern red muntjac. Mêdog County, Tibet AR. 2014. Photo by Li Cheng.
138	Forefoot track of northern red muntjac. Note the two hooves are well separated. Xishuangbanna NR, Yunnan Province. 2008. Photo by Sun Ge.
139	One pair of Williamson's chevrotains in a rainforest. Xishuangbanna NR, Yunnan Province. 2008. Photo by Feng Limin.
140	Adult female alpine musk deer (tusks absent) in the rocky alpine habitat during summer. Sanjiangyuan NR, Qinghai Province. 2012. Photo by Dong Lei/IBE.
141	Adult male alpine musk deer in the open alpine shrub habitat during winter. Sanjiangyuan NR, Qinghai Province. 2014. Photo by Peng Jiansheng/IBE.
142	Latrine site of alpine musk deer with scat at varied ages. Shenxianshan NR, Sichuan Province. 2016. Photo by Li Sheng.
143	Adult male of forest musk deer. Foping NR, Shaanxi Province. 2008. Photo by Yong Yange.
144	Track of forest musk deer on snow. Note the distinct dew claws. Changqing NR, Shaanxi Province. 2011. Photo by Xiang Dingqian.
144	Hairs of forest musk deer left on a leaning conifer tree. Shenxianshan NR, Sichuan Province. 2006. Photo by Li Sheng.
145	Adult female red deer in summer coat. Guoluo Prefecture, Qinghai Province. 2013. Photo by Wu Lan.
146	Adult male red deer. Shannan County, Tibet AR. 2014. Photo by Guo Liang/IBE.
147	Adult male sambar. Wolong NR, Sichuan Province. 2016. Photo by PKU Wildlife Lab.
148	Sambar track of hind foot (with dew claw on left). Anzihe NR, Sichuan Province. 2015. Photo by Fu Qiang.
149	Adult female sambar. Although typically found in warm and moist forest, sambars at their northern extent in Sichuan Province are also found at high elevation. Wolong NR, Sichuan Province. 2015. Photo by PKU Wildlife Lab.
150	Adult male Siberian roe deer during the fall. Taihang Mountains, Shanxi Province. 2013. Photo by Song Dazhao.

Page	Image
170	Regular round fecal pellets of blue sheep. Sanjiangyuan NR, Qinghai Province. 2011. Photo by Wu Lan.
171	Dark pelage form of Chinese goral. Changqing NR, Shaanxi Province. 2004. Photo by Xiang Dingqian.
171	Pale gray pelage form of Chinese goral. Changqing NR, Shaanxi Province. 2010. Photo by PKU/eMammal.
172	Brownish pelage form of Chinese goral. Tangjiahe NR, Sichuan Province. 2007. Photo by Dong Lei.
172	Regular round fecal pellets of Chinese goral (upper pile) in comparison to those of Chinese serow (lower pile), which are larger and longer. Jiuzhaigou NR, Sichuan Province. 2014. Photo by Li Sheng.
172	Track of Chinese goral left on wet sandy ground. Laohegou NR, Sichuan Province. 2012. Photo by Li Sheng.
173	Adult male Himalayan goral. Zhangmu, Tibet AR. 2006. Photo by Dong Lei/IBE.
174	Adult red goral. Yaluzangbu NR, Tibet AR. 2014. Photo by Wu Xiushan/IBE.
175	Fresh fecal pellets of red goral left on rock surface. Mêdog County, Tibet AR. 2014. Photo by Li Cheng.
175	Adult red goral. Yaluzangbu NR, Tibet AR. 2014. Photo by Wu Xiushan/IBE.
176	Adult Chinese serow. Wolong NR, Sichuan Province. 2010. Photo by PKU/eMammal.
177	Large piles of Chinese serow fecal pellets at different ages, varying from fresh (lower right pile) to old (upper pile). Jiuzhaigou NR, Sichuan Province. 2014. Photo by Li Sheng.
178	Adult Himalayan serow. Yaluzangbu NR, Tibet AR. 2010. Photo by Dong Lei/IBE.
179	Adult male Przewalski's gazelle. Note the horn tips are pointing toward each other. Qinghaihu NR, Qinghai Province. 2012. Photo by Bu Hongliang.
179	Juvenile female Przewalski's gazelle. Qinghaihu NR, Qinghai Province. 2009. Photo by Liu Jiazi.
180	Fresh track of Przewalski's gazelles in typical habitat of steppe grassland during the summer growing season near the Qinghai Lake. Qinghaihu NR, Qinghai Province. 2008. Photo by Liu Jiazi.
181	Typical and unique habitat of Przewalski's gazelles around the Qinghai Lake during late fall. Qinghaihu NR, Qinghai Province. 2007. Photo by Wang Dajun.
181	Fresh fecal pellets of Przewalski's gazelle. Qinghaihu NR, Qinghai Province. 2008. Photo by Liu Jiazi.
181	Scratching pits and scats left by Przewalski's gazelle during the mating season in winter. Qinghaihu NR, Qinghai Province. 2012. Photo by Bu Hongliang.
182	Forefoot print of Przewalski's gazelle. Qinghaihu NR, Qinghai Province. 2008. Photo by Liu Jiazi.
182	Resting beds left by Przewalski's gazelles in the dry grasses during winter. Qinghaihu NR, Qinghai Province. 2010. Photo by Liu Jiazi.
182	Adult male Tibetan gazelle in summer coat. Sanjiangyuan NR, Qinghai Province. 2012. Photo by Dong Lei/IBE.
183	Adult female Tibetan gazelles in their winter coat. Sanjiangyuan NR, Qinghai Province. 2013. Photo by Wu Lan.
184	Adult male Tibetan gazelle in winter coat. Note the upper parts of the horns are almost parallel to each other. Sanjiangyuan NR, Qinghai Province. 2013. Photo by Wu Lan.
184	Old dry fecal pellets of Tibetan gazelle left on open alpine grassland. Yajiang County, Sichuan Province. 2016. Photo by Li Sheng.
185	Adult male Tibetan antelope during winter. Sanjiangyuan NR, Qinghai Province. 2008. Photo by Xiang Dingqian.

Page	Image
186	Adult male Tibetan antelope during summer. Qiangtang NR, Tibet AR. 2016. Photo by Luo Chunping.
186	Juvenile male Tibetan antelope during summer. Sanjiangyuan NR, Qinghai Province. 2012. Photo by Dong Lei/IBE.
187	Tibetan antelope track and scats left on the sandy ground. Qiangtang NR, Tibet AR. 2015. Photo by Liang Xuchang.
187	Close-up view of Tibetan antelope fecal pellets. Qiangtang NR, Tibet AR. 2015. Photo by Liang Xuchang.
188	Adult gaur foraging in the tropical forest. Nabanhe NR, Yunnan Province. 2015. Photo by Cao Guanghong.
189	Adult gayal (*B. gaurus frontalis*). The horns are more open and less curved pointing outward than those of the gaur. Mêdog County, Tibet AR. 2015. Photo by Cheng Bin/IBE.
190	Hind foot track of gaur. Nabanhe NR, Yunnan Province. 2016. Photo by Cao Guanghong.
190	Fresh scat of gaur as a single large mass. Nabanhe NR, Yunnan Province. 2010. Photo by Cao Guanghong.
190	Old, flat, pie-shaped scat of gaur. Nabanhe NR, Yunnan Province. 2012. Photo by Cao Guanghong.
191	Young adult female golden takin (*B. t. bedfordi*). Changqing NR, Shaanxi Province. 2008. Photo by PKU/eMammal.
192	Adult male golden takin (*B. t. bedfordi*) during the mating season in July. Changqing NR, Shaanxi Province. 2008. Photo by PKU/eMammal.
192	Adult Bhutan takin (*B. t. whitei*). Yaluzangbu NR, Tibet AR. 2011. Photo by Dong Lei/IBE.
193	Two adult female Sichuan takin (*B. t. tibetana*) with a calf. Tangjiahe NR, Sichuan Province. 2004. Photo by Li Sheng.
194	Fresh separated fecal pellets of takin. Huangbaiyuan NR, Shaanxi Province. 2010. Photo by Li Sheng.
194	Dry clustered scat of takin. Changqing NR, Shaanxi Province. 2010. Photo by Li Sheng.
194	Fresh loose patty scat left by takin while foraging on young grass. Laohegou NR, Sichuan Province. 2012. Photo by Li Sheng.
194	Track of takin. Laohegou NR, Sichuan Province. 2012. Photo by Li Sheng.
195	Takin feeding sign on tree bark in spring. Laohegou NR, Sichuan Province. 2006. Photo by Li Sheng.
195	Rub marks on conifer tree. Wanglang NR, Sichuan Province. 2004. Photo by Li Sheng.
195	Takin hairs left on tree stem after a takin rubbed against the tree. Tangjiahe NR, Sichuan Province. 2013. Photo by Li Sheng.
195	Mineral soil used by takin; note the distinct teeth mark. Tangjiahe NR, Sichuan Province. 2006. Photo by Li Sheng.
196	Mother-foal group of kiang during summer. Sanjiangyuan NR, Qinghai Province. 2011. Photo by Wu Lan.
197	Kiang scat pile. Sanjiangyuan NR, Qinghai Province. 2011. Photo by Wu Lan.
197	Dry, oval, flat fecal pellets of kiang. Qiangtang NR, Tibet AR. 2015. Photo by Liang Xuchang.
198	Large foraging herd of kiang in typical open alpine meadow habitat. Sanjiangyuan NR, Qinghai Province. 2011. Photo by Wu Lan.
198	Adult male wild boar. Tangjiahe NR, Sichuan Province. 2004. Photo by PKU/eMammal.
199	Fresh scat left by wild boar. Laohegou NR, Sichuan Province. 2012. Photo by Li Sheng.

Page	Image
199	Rare dry fecal pellets left by wild boar that are tied together by long grass fibers. Jiuzhaigou NR, Sichuan Province. 2014. Photo by Li Sheng.
200	Fresh tracks of wild boar left on soft soil. Note the hind hoof (bottom) imprinted over the fore hoof (top). Laohegou NR, Sichuan Province. 2012. Photo by Li Sheng.
200	Digging pit left by wild boars on open meadow. Wanglang NR, Sichuan Province. 2011. Photo by Li Sheng.
200	Digging pit left by wild boar while feeding on tree root. Laohegou NR, Sichuan Province. 2006. Photo by Li Sheng.
200	Bedding site made of arrow bamboo by wild boar in conifer forest. Wanglang NR, Sichuan Province. 2004. Photo by Li Sheng.
200	Adult wild boar visiting a mud wallow. Changqing NR, Shaanxi Province. 2008. Photo by Li Sheng.
201	Pangolin photographed on Taiwan. 2011. Photo by Huang Yifeng.
202	Chinese pangolin burrow. Shangyong Nature Reserve, Yunnan Province. 2014. Photo by Sun Ge.
202	Chinese pangolin burrow. Shangyong Nature Reserve, Yunnan Province. 2014. Photo by Sun Ge.
203	Asiatic brush-tailed porcupine. Gaoligongshan NR, Yunnan Province. 2016. Photo by KFBG-KCC.
204	Adult Malayan porcupine. Wanglang NR, Sichuan Province. 2008. Photo by Li Sheng.
205	Malayan porcupine habitat in secondary deciduous forest; the porcupine usually follows the ridge trail to forage. Niuweihe NR, Shaanxi Province. 2010. Photo by Wang Fang.
205	Malayan porcupine scat. Wanglang NR, Sichuan Province. 2006. Photo by Li Sheng.
206	Malayan porcupine scat pile. Tangjiahe NR, Sichuan Province. 2015. Photo by Li Sheng.
206	Malayan porcupine quill on the ground. Tangjiahe NR, Sichuan Province. 2015. Photo by Li Sheng.
206	Feeding site of the Malayan porcupine; note the bare trunk on the front side of the trunk. Changqing NR, Shaanxi Province. 2013. Photo by Xiang Dingqian.
207	Black giant squirrel. Gaoligongshan NR, Yunnan Province. 2014. Photo by Li Jiahong.
207	Black giant squirrel feeding site. Gaoligongshan NR, Yunnan Province. 2015. Photo by Li Jiahong.
208	Himalayan marmot. Sanjiangyuan NR, Qinghai Province. 2011. Photo by Wu Lan.
209	Typical Himalayan marmot habitat of alpine meadow and mountains. Sanjiangyuan NR, Qinghai Province. 2010. Photo by Wang Fang.
209	Himalayan marmot track. Sanjiangyuan NR, Qinghai Province. 2016. Photo by Tashi Sanger.
210	Himalayan marmot within an alpine scree slope. Wolong NR, Sichuan Province. 2010. Photo by Li Sheng.
210	Himalayan marmot scat; note the beetle shells. Yajiang County, Sichuan Province. 2006. Photo by Li Sheng.
210	Entrance of a Himalayan marmot burrow. Luhuo County, Sichuan Province. 2006. Photo by Li Sheng.
210	Dense Himalayan marmot burrows on a slope of alpine grassland. Yajiang County, Sichuan Province. 2016. Photo by Li Sheng.
211	Yunnan hare habitat. Yaoshan NR, Yunnan Province. 2016. Photo by Wang Fang.
212	Yunnan hare scat. Yaoshan NR, Yunnan Province. 2016. Photo by Wang Fang.

Page	Image

Page	Image

Page	Image
274	Female Tibetan snowcock on her nest. Lhasa City, Tibetan AR. 2004. Photo by Lu Xin.
274	Nest and eggs of Tibetan snowcock. Lhasa City, Tibetan AR. 2004. Photo by Lu Xin.
276	Domestic dog (Tibetan mastiff) scats after foraging yak carcass. Yajiang County, Sichuan Province. 2016. Photo by Li Sheng.
276	Domestic dog tracks; note the claw marks. Wolong NR, Sichuan Province. 2006. Photo by Li Sheng.
277	Domestic cattle scats; note the clumped large fecal pellets. Wanglang NR, Sichuan Province. 2016. Photo by Li Sheng.
277	Domestic cattle scats; note the flat and structureless single pile. Wanglang NR, Sichuan Province. 2016. Photo by Li Sheng.
277	Domestic water buffalo tracks. Chongzuo NR, Guangxi AR. 2015. Photo by Li Sheng.
278	Domestic goat scat. Qinghaihu NR, Qinghai Province. Photo by Wang Fang.
279	Domestic horse scats. Wanglang NR, Sichuan Province. 2016. Photo by Li Sheng.
279	Domestic horse tracks; note the single-hoof prints. Wanglang NR, Sichuan Province. 2016. Photo by Li Sheng.
280	Bat droppings. Laohegou NR, Sichuan Province. 2012. Photo by Li Sheng.
280	A moupin pika (*Ochotona thibetana*) in its typical alpine open habitat. Wolong NR, Sichuan Province. 2009. Photo by Li Sheng.
280	Burrow and latrine site of moupin pika in alpine meadow. Wanglang NR, Sichuan Province. 2012. Photo by Li Sheng.
281	The most commonly encountered flying squirrel species in southwestern China, the gray-headed flying squirrel (*Petaurista caniceps*). Changqing NR, Shaanxi Province. 2012. Photo by Xiang Dingqian.
281	Large collection of walnut shells left by squirrels under a rock. Tangjiahe NR, Sichuan Province. 2006. Photo by Li Sheng.
281	Broken branches with missing bark and typical round fecal pellets left by the flying squirrel *Trogopterus xanthipes*. Jiuzhaigou NR, Sichuan Province. 2014. Photo by Li Sheng.
281	Walnut shells left by Pere David's rock squirrel (*Sciurotamias davidianus*). Tangjiahe NR, Sichuan Province. 2006. Photo by Li Sheng.
282	Feeding site of Chinese bamboo rat (*Rhizomys sinensis*), the most common bamboo rat species within the region. Note the bamboo stem to the left was cut, and the soil to its right was disturbed as the bamboo rat came out from its underground burrow. Changqing NR, Shaanxi Province. 2016. Photo by Xiang Dingqian.
282	Lesser bamboo rat (*Cannomys badius*) at its burrow entrance. Tongbiguan NR, Yunnan Province. 2009. Photo by Li Sheng.
283	Pellets produced by the Eurasian eagle owl (*Bubo bubo*). Beijing. 2014. Photo by Song Dazhao.
283	Hole nests of blue-tailed bee-eater *Merops philippinus* on a vertical river bank. Baoshan City, Yunnan Province. 2006. Photo by Li Sheng.
283	Holes left by three-toed woodpecker on a large conifer tree. Wanglang NR, Sichuan Province. 2013. Photo by Li Sheng.

INDEX OF COMMON NAMES

INDEX OF SPECIES NAMES

All species are mentioned in text. Species with entries in this guide are in **bold**.